CHANGING JOBS

THE FAIR GO IN THE NEW MACHINE AGE

Jim Chalmers
& Mike Quigley

Published by Redback Quarterly,
an imprint of Schwartz Publishing Pty Ltd
Level 1, 221 Drummond Street
Carlton VIC 3053, Australia
enquiries@blackincbooks.com
www.blackincbooks.com.au

National Library of Australia Cataloguing-in-Publication entry:
Mike Quigley, author.
Changing jobs: the fair go in the new machine age/
Mike Quigley; Jim Chalmers.
9781863959445 (paperback)
9781925435894 (ebook)
Industrial sociology – Australia.
Job satisfaction – Australia.
Information society – Social aspects – Australia.
Self-actualization (Psychology).
Other Creators/Contributors: Chalmers, Jim, author

Cover design by Peter Long
Cover image by Scanrail1, Shutterstock
Typesetting by Tristan Main

Printed in Australia by McPherson's Printing Group.

For Vicki, Siobhan, Roisin and Aislinn, who saved an engineer from grammatical embarrassment, and Laura, Leo and little Annabel, who arrived while the book was being written.

CONTENTS

PREFACE

There is a tremendous upside to technological change. It has the potential to improve lives and wellbeing, save time and effort and help combat, if not overcome, so many of the obstacles to a good life in a thriving society. But agreeing that technological change can improve living standards does not mean dismissing the real fears that people have about where, or *whether*, they fit in a workforce increasingly dominated by machines.

This book covers a lot of ground but principally it's about technology and the changing face of work, and how best to approach this change – at both a personal and a political level – to advance the fair go in the new machine age.

One of the authors is a technologist and former chief executive officer with a telecommunications background; the other is a politician with experience in economic policy and a PhD in political science. One devours *Scientific*

American, the other the *Financial Times*. One lives in the centre of Sydney, the other is from Logan, south of Brisbane. One is a baby boomer, the other Gen X.

In writing this book together about technology and its impact on jobs in Australia, we both accept that technology will help us produce more goods and services, but as progressive people we also care deeply about how the economic gains are distributed. We believe that accelerating inequality is already a serious problem; that it must be addressed; and that *it is not inevitable* – that bursts in technology need not result in bursts of inequality.

We draw inspiration from the American economist John Kenneth Galbraith, who wrote four decades ago that 'all of the great leaders have had one characteristic in common: it was the willingness to confront unequivocally the major anxiety of their people in their time'.[1] For many Australians, precarious and insecure work and the rapid growth in artificial intelligence, robotics, automation and machine learning are the defining and dominant anxieties they face today. What will this mean for their jobs, now and in years to come? What will it mean for their kids' futures? This anxiety feeds discontent with the political system, which in turn fuels a search for extreme, simplistic and backward-looking 'solutions'.

Michael Ignatieff, the one-time Canadian Opposition leader, wrote recently of the historical role of figures such as William Gladstone and Teddy Roosevelt. They understood that political leaders should not resist the forces of change, but instead bring in 'the excluded and hold ... in check the power of the barons enriched by the new technologies'.[2]

The point is, we have choices. But leadership and foresight are required. This begins with understanding the changes and caring enough to act in the interests of those in the middle who are squeezed, as well as those at risk of falling behind. It means understanding the shifts we know are coming, as well as examining those that may seem far-fetched but are nevertheless worth preparing for, and recognising that many of them will come from abroad, impacting on our domestic choices. It means democratic institutions, workplace arrangements and decision-making processes that allow society to cope with rapid and sometimes unexpected change.

Perhaps, above all else, it means accepting and understanding that this revolution is different. While earlier revolutions replaced human effort, this one goes a step further to directly challenge some intrinsic traits that make us human – thinking, problem-solving and decision-making.

What does this mean for Australians' jobs?

Do automation and artificial intelligence represent a revolution or an evolution of the workplace? Are we seeing a step-change or something more incremental? Will we see more jobs destroyed, or created, or simply gradually changed? We will explore these questions, but much of what we propose applies in a 'no regrets' way – it will be helpful whatever the future brings.

The starting point of the book is the expectation that technological change could – if left unattended – worsen already-worrying trends towards inequality. That's because these changes risk further skewing power relationships for ordinary people at work, with troubling consequences for wages and conditions.

Our point is that there is no such thing as technological trickle-down. As with economic gains more broadly, those achieved by artificial intelligence, automation, machine learning and robotics will not simply share themselves. Even when machines bring us new sources of prosperity and wealth, there is no guarantee that we will all benefit together. If Australia still cherishes the fair go, like the authors do, then now is the time to think carefully about the changes we need to see in the decade ahead.

There are broadly three paths to choose from. One is taken by the 'let it rip' crowd, who cheer on technological change without regard for wealth concentration or impacts on real people. This group believes, as Malcolm Turnbull does, that these are 'exciting times', regardless of the consequences for those disrupted. Another group argues we can resist technological change or hold it back. This is about as likely as offices rediscovering a preference for the fax machine.

The third group believes in intelligent, meaningful interventions, correcting market failures, investing in lifelong learning, rethinking industrial relations, and renewing and restitching the social safety net. This path is the one we recommend for Australia. What differentiates our approach is the conviction that we can attack the worst consequences of technological change without denying ourselves the broader benefits.

To a large extent, Australians will choose the level of inequality we are prepared to tolerate in this country, consistent with our values. The task is to capitalise on national strengths such as our adaptability, the targeted nature of our welfare system, a quarter-century of continuous economic growth, and more. Our principles translate into three objectives: growth that is inclusive; work that is rewarded; and a decent safety net for those left behind, including support to help them transition to meaningful work if possible.

But Australia, like the rest of the developed world, is dangerously ill-prepared for the scenarios we will discuss – either a serious decline in the total number of jobs or hours, or, perhaps more likely, a step-change in the mix of jobs and the nature of work. Our teaching and training institutions, industrial relations regime and social security will all need to change as the rules of the global and domestic economies are rewritten by machines and those who own them. We shouldn't kid ourselves that technological change in the workplace can go unaccompanied by changes in our schools, governments and homes.

This book is for everyone who cares about this as much as this unorthodox combination of authors does. Neither policy-makers nor technologists have the answers on their own. Rapid technological transformations are already changing our workforce and society in ways that threaten the fair go we cherish in this country. We need collaboration, anticipation, planning and a change of mindset at the personal level to ensure Australians are beneficiaries – not victims – of this change.

CHAPTER 1
AN INTRODUCTION TO THE NEW MACHINE AGE

Mike has been using Duolingo, a language learning application launched in 2011. It is very effective at teaching even a novice a new language. It is fun to use, completely free, asks for no donations and, unlike many free websites, subjects the user to minimal advertising.

What was the business model behind this incredibly useful application?

Duolingo was developed by Luis von Ahn, a professor of Computer Science at Carnegie Mellon University in Pittsburgh, along with a small team that included one of his graduate students, Severin Hacker – a great name for a website developer. They had asked themselves the question: 'How could you get 100 million people to translate the web into every major language for free?' Even if there were enough translators, it would take a lot of money and time. They estimated that just translating Wikipedia into Spanish – in fact, only the part that hadn't already been translated – would cost at least US$50 million if professional translators were employed.

The solution they developed became Duolingo.

The idea was to provide free language courses to large numbers of people and, as part of the course, each student would translate sentences from Wikipedia. Of course, a beginner student's translations would likely contain errors. However, testing showed that if the translations of a number of beginners were combined using an automated algorithm, the resulting translation could be made as accurate as that provided by a professional translator.

This is not only a clever idea, but also seems to be very equitable – the students gain access to a totally free, high-quality language course, and in return provide their time and effort. It is a win for everyone – or almost everyone. If you are a professional translator, language teacher or someone who has developed online language courses you are trying to sell, this model may not look so attractive. But notwithstanding that downside, Duolingo is a valuable initiative that is bringing language lessons to many people, including those who otherwise could not afford them.

This is just one example of the rapid advances taking place in technology. Some of them are likely to have a major impact on the employment prospects, and hence the potential prosperity and wellbeing, of our children and grandchildren. But the suggestion that we should try to halt or even delay the advance of technology is neither practical nor desirable.

We both believe that technology has been a huge boon to humanity, and will continue to be so for the foreseeable future. Benefits include national electrical grids to power our increasing array of machines, global communications networks that allow us to communicate almost instantly, high-speed transport systems, antibiotics, vaccines and sophisticated diagnostic tools that have seen steady increases in lifespans in most of the Western world. The list of life-changing technologies continues to grow.

There are also less dramatic but still important benefits. Technological tools have made the completion of many routine and sometimes tedious tasks more efficient, saving people considerable amounts of time, and giving them more options when it comes to organising their lives, including their work lives.

We believe that the technological advances we are going to see in the coming decades, in areas such as computation, networking, artificial intelligence (AI), analytics and robotics, will change the nature of work in Australia and compel Australians and their governments to find fair and equitable solutions to what may be a challenging employment environment.

Although precise predictions about systems as complex as our economy or the evolution of our social structures are not possible, it can still be useful to consider some of the broader trends that may occur over time. To draw an analogy with another complex system, it is not possible to make accurate predictions about the weather more than a week or two ahead. But it is possible, and indeed essential, to think about the long-term trends in climate.

Continuing the comparison a little further, governments are generally effective at responding to immediate threats due to severe events such as hurricanes. But it has been much more difficult for governments to plot an agreed course in tackling the slowly (but surely) developing problem with climate warming. Disappointingly, it has been difficult to reach a political consensus that the problem even exists.

The same is true regarding the impact of technology on future employment in Australia. There is no general consensus among economists, industry experts and technologists that advancing technology is going to result in significant reductions in the workforce. Some experts insist they have heard this story before and point to the agricultural, industrial and computer revolutions, where many jobs were destroyed but many more were created. They believe it will be the same with automation, networking, robotics and AI: while jobs may be substituted by rapidly improving technologies, these same technologies will generate new, but different jobs.

This was a theme identified in a 2014 Pew Research Center report.[1] Of the nearly 2000 experts consulted, 52 per cent thought that by 2025 technology would displace certain types of work but, largely because this is what happened in the past, they thought technology would have a neutral to positive impact on jobs. It is certainly true that the displacement of agricultural workers by agricultural machinery, skilled artisans by unskilled production workers, and production workers by automated manufacturing systems – along with the reduction in routine office work by IT systems – did not result in an overall reduction in employment in the long term.

However, other experts, including 48 per cent of the respondents cited in the Pew report, take a more pessimistic view and believe that 'this time it is different'. They argue we are now seeing the start of a trend in job replacement by technology, particularly robotics and artificial intelligence, that is not going to be reversed. If this trend continues, it will lead to an increasing percentage of workers being displaced. And as we will see, it won't just be low-income and low-skilled jobs at risk. Mid-tier jobs are also increasingly under threat and, perhaps even more surprisingly, many high-skilled and high-income jobs aren't likely to be immune either. The flow-on effect from these jobs being made redundant is a lack of workers' disposable income to create demand elsewhere in the economy.

We both believe that the continuing advances in technology will cause some significant disruptions in employment in Australia, and many jobs of the past may disappear. A prudent approach for policymakers would therefore be to anticipate and plan for these disruptions.

Our aim with this book is to propose some robust initiatives and interventions that will have no significant downsides if the impact of technology turns out to be relatively benign or indeed positive. In other words, we aim to identify a 'no-regrets' strategy.

Being prepared means not only being in a position to take advantage of the technologies, but also putting in place contingencies to prevent a large number of Australians from being left in a chronically disadvantaged position. Large numbers of people being left behind for an extended period is not conducive to social stability.

Markets solve many problems and have helped create the wealth and prosperity that underpin our Australian living standards. But markets alone are not always effective in solving broad-ranging and complex problems without the considered and careful intervention of government. Most of us would not want to see either health or education left entirely to the market; we understand that governments have an important role to play in these areas. Most people also understand what can happen when inadequate regulation is taken advantage of by those in a position to do so – they have seen the outcome with the GFC and, more recently, with the energy sector.

Our task is to do what we can to ensure markets and advancing technologies work for people and not against them, and to avoid a bifurcation of the labour market that would see some people assisted at work by intelligent machines and others replaced by them altogether.

The challenge for government is to encourage the adoption and use of new technologies, and to ensure that Australia's ability to innovate and build new businesses is not diminished, while also promoting equality.

THE MARCH OF TECHNOLOGY

Computation and digital networking have been advancing rapidly during the last few decades and there is no sign that this trend is slowing down. In fact, there have been some dramatic advances in computational power in the last few decades.[2] In 1996, the first supercomputer made from off-the-shelf central processing units (CPUs) was created. It was called ASCI Red and contained more than 9000 processors, cost almost US$50 million to build and was the first supercomputer to break the 1 teraflop (computing speed) barrier. ASCI Red occupied an entire room and consumed almost one megawatt of power. In 2011, just fifteen years later, Intel announced that they had developed a single chip that could operate at 1 teraflop.[3]

This increasing computational power has been one of the factors, together with improvements in algorithms, that has helped solve some of the challenges faced by robotics and artificial intelligence researchers. As a consequence, machines are now able to progressively improve their own performance in areas such as speech recognition and language processing, vision and image recognition and inference.

How far can AI technology progress within the next few decades?

Nick Bostrom, the founding director of both the Future of Humanity Institute and the Programme on the Impacts of Future Technology at Oxford University, provides an indication in his recent book.[4] Based on a number of expert surveys and interviews, he writes that 'it may be reasonable to believe that human-level machine intelligence has a fairly sizeable chance of being developed by mid-century'. He defines human-level machine intelligence as 'one that can carry out most human professions at least as well as a typical human'.

Bostrom emphasises that these types of predictions are always subject to large uncertainties, and that human-level machine intelligence may be reached much later than mid-century – or much earlier. Bostrom is wise to highlight the risks in predicting technological advances. These advances are often interdependent in ways that are very difficult to anticipate. For example, it would have been hard to predict that mobile phones were going to impact the camera market in such a profound way. There is also the possibility that a future technology falls well short of what was expected of it. Commercial supersonic flight is one example of this.

Bostrom's prediction regarding the timing of human-level machine intelligence is supported by some impressive recent feats in AI. In 2011 IBM's 'Watson' – a computer system designed specifically for answering questions in natural language – beat the two reigning champions in the general knowledge quiz show *Jeopardy!* In 2012 a prediction was made that AI might beat the human world champion in the complex 3000-year-old game Go in about a decade – in 2022. But in March 2016, just four years after that prediction, the AI program AlphaGo (developed by the Google-owned British company DeepMind) beat then-world champion Lee Sedol. The AlphaGo algorithm 'retired' in early 2017, having swept the world number one, nineteen-year-old prodigy Ke Jei, in a three-game series. It will now focus on complex challenges such as 'finding new cures for diseases, dramatically reducing energy consumption, or inventing revolutionary new materials'.[5]

Another much-quoted example of advancing AI is the self-driving car, being pursued by Tesla, Google and many others. The AI systems being used in the trials of these cars include sensors of various sorts, combined with considerable processing power. They not only detect cars around them but also calculate the 'blind spots' of those cars. But these cars cannot, at least not yet, deal with all the unanticipated events that occur on typical suburban roads the way a human driver does. Dealing with snow, different types of road boundaries and T-intersections are all proving to be challenging.

Some artificial intelligence researchers are paying increasing attention to the disciplines of neuroscience and psychology to better understand how human minds solve problems. A two- or three-year-old human can rapidly learn tasks with relatively little input, whereas the best AI systems today require vast amounts of data in order to learn the same tasks.

While AI and robotics currently have some very real limitations, they will undoubtedly become more capable as computation, algorithms, sensors and actuators continue to improve. Those capabilities will be enhanced by the ability to tap into massive amounts of data available almost everywhere due to the growing ubiquity of digital connectivity and the dramatically reduced cost of data storage.

Everyone has heard the stories of routine factory jobs being replaced by machine automation and industrial robots, or drivers in remote mine sites being replaced by driverless vehicles. What is probably not as well appreciated is that the impact is likely to extend to more highly skilled jobs. Duolingo is still only in its very early days, and one can imagine how much more useful it will become with the increasing application of advanced artificial intelligence. As it 'learns' from the immense number of sentences being translated by human students, the need for human input for the translation task will likely decrease. The jobs it will be capable of replacing – professional translator and language teacher – would not be regarded today as low-skilled occupations.

Of course, it can be argued that Duolingo is not replacing human translators or language teachers, but just adding to the amount of translation and language tuition that can be done. But is it really likely that these jobs won't be put at risk by the existence of a sophisticated translation and teaching application that is low-cost, available at any time and can work all day, every day, without a break?

Many manufacturing jobs have already been replaced by robotics and automation, while many clerical tasks have been replaced by information technology. This trend will continue as robotics and verbal and image processing become increasingly capable. So the jobs of clerks, assessors, appraisers and agents working in a variety of fields – real estate, insurance, law and transport, to name just a few – are at risk of being replaced by machines.

The skilled professions are also not protected. Radiologists may initially see machine interpretation of radiological images as a useful tool that will assist them with their work. But in the longer term, it is possible that machine intelligence will be able to interpret radiological images more accurately, much faster and for a fraction of the cost of a human radiologist. Computer interpretation of image data is one of the areas where advances in machine capability are occurring rapidly.[6] In a related area, the growing capability of AI to perform medical diagnosis may result in medical training focusing more on communication and accurately establishing symptoms and less on diagnostic skills.

The jobs that are not so easily displaced by robots and

AI are the non-routine jobs that require a high level of physical dexterity, situational awareness or intensive human interaction, or those requiring high levels of creativity, analytical skills or emotional intelligence. This means that occupations such as dentists, therapists, human resources managers and ministers of religion are less likely to be replaced anytime soon – although those holding these occupations are likely to see more sophisticated and intelligent tools becoming available to assist them with their work. Indeed, it is a very plausible scenario that what we will see in the decades ahead is the widespread provision of AI and robotics tools to assist people in doing their work.

The Australian economy has, over a long period, been transitioning away from being largely based on agricultural and industrial activities and towards service industries. It could be thought that this will shield the Australian economy, more so than other economies, from the impacts of advancing technology. Unfortunately this is a vain hope, for two reasons. First, Australia has not made the transition to a largely service-based economy any faster than most other Western countries. In fact the US, UK and German economies all have a greater proportion of GDP based on service industries than does Australia. Australia is on a par with Canada in terms of the proportion of its economy that is service-based.

Secondly, while some jobs in the service industries are less likely to be easily replaceable by automation and computerisation, it is by no means true for all types of service jobs. As mentioned previously, service jobs in the areas of law, insurance and real estate are quite susceptible to being automated.

Some jobs requiring what we would describe as high intellectual input, such as a translator, are at risk, but jobs requiring low levels of intellectual input together with high levels of dexterity are safe. This is an illustration of Moravec's paradox. In the late 1980s, Hans Moravec noted that high-level reasoning requires relatively little computational power, but simulating low-level sensorimotor skills requires considerable computational resources.

Hence a chess-playing computer beat the world's chess champion, Garry Kasparov, back in 1997, but a robot equipped with the most sophisticated AI could not, today, play a game of tennis.

Steven Pinker, the Harvard cognitive scientist, put it this way:

> The main lesson of 35 years of AI research is that hard problems are easy and easy problems are hard. As the new generation of intelligent devices appears, it will be the stock analysts, petrochemical engineers and parole board members who are in danger of being replaced by machines. The gardeners, receptionists and cooks are secure in their jobs for decades to come.[7]

'Job polarisation' has been a trend in labour markets for some time. It is characterised by a reduction in the demand for middle-income jobs and a relative increase in demand for both low-income and high-income jobs. The powerful processing platforms available today and the intelligent machines that are being developed excel at the routine processing of information or the routine analysis of data at a speed and accuracy no human can match. Hence machines are well placed to perform the type of routine clerical or administrative work that has until now been rewarded with a mid-level income. Likewise, robotics and automation have replaced many of the mid-level jobs in factories, mines and other industrial workplaces.

Demand is seen at both ends of the income spectrum – for senior executives, statisticians and systems analysts at one end and gardeners, hairdressers and baristas at the other.

A much-quoted 2013 paper written by Oxford University researchers Carl Frey and Michael Osborne provides a detailed analysis of more than 700 occupations.[8] Based on US data, the paper attempts to estimate how susceptible each occupation is to automation via computer-controlled equipment. The conclusions are startling: nearly half of all (US) jobs are at risk within a decade or two, and the researchers say their findings also apply to other developed economies.

A number of subsequent studies have built on this work, two of which looked specifically at Australian data. The conclusions they drew were broadly similar.[9] A number of other studies, however, suggest the methodology Frey and Osborne used leads to overly pessimistic predictions.

EMERGING JOBS

A question most of these studies do not try to answer is the number of new jobs likely to be created by advances in technology. Unfortunately, the data regarding the emergence of new jobs is not encouraging.

As at 2010, only half of 1 per cent of all jobs in the US are in industries that came into existence after the year 2000.[10] So in ten years, these new industries had very little impact on the number of jobs created. The instant-messaging software company WhatsApp, for example, had only fifty-five employees when Facebook bought it for US$19 billion. This seems to be the nature of these digital businesses – they do not require a lot of capital investment or many employees, but they create great wealth for those who generate the innovation or have the wherewithal and foresight to invest in that innovation.

It is, of course, the case that only a small percentage of digital start-ups are as successful as the businesses mentioned above. Most start-ups fail. This is why the competitive dynamics of these new digital businesses is often described as 'winner takes all'.

There is another trend at work also, as American writer Jaron Lanier observes.[11] Not only do many of these more recent networking businesses make a small number of individuals very rich, they do so by relying on the unpaid inputs of a large number of people. Take, for example, LinkedIn, a social networking site providing services to both individuals and organisations. At the beginning of 2016, there were more than 400 million registered members and more than 100 million active users, over 200 countries. The value to the individual users comes from their ability to create and manage an online professional profile, build a network of professional contacts and have access to job opportunities posted on the LinkedIn platform. The value that accretes to LinkedIn comes from the sale of information contained in the user-created profiles and the ability to advertise to the large user base. Given that LinkedIn was acquired by Microsoft at the end of 2016 for approximately US$26 billion, considerable value could clearly be leveraged from this very large base of user-provided information.

Digital and networking technologies are already having an effect on the historical patterns of income and wealth distribution. This is seen most starkly in the US. Real income of the median US household fell by nearly 10 per cent from 1999 to 2011, and the top 1 per cent of US earners took 22 per cent of total income. The top 20 per cent of US individuals received more than the total increase in wealth from 1983 to 2009 because the wealth of the bottom 80 per cent decreased during this period.[12]

The situation in other OECD countries is not as extreme as in the US, but the trends are in the same direction.

A 2017 OECD economic survey found that, over a decade, incomes for the poorest 20 per cent of Australians rose by just over 25 per cent, but increased more than 42 per cent for the richest fifth of Australians. More concerningly, the average net wealth of an Australian in the bottom 20 per cent only increased 4 per cent during the same period, compared with a 38 per cent increase for the top 20 per cent.[13] The reasons for this growing inequality in Australia are thought to include globalisation, reductions in the top tax rates, increasing use of tax exemptions and the displacement of jobs due to technology.

THE ROLE OF EDUCATION

It is often suggested that the way to reverse these employment trends is to put more emphasis on education and training – educating our younger people in those areas likely to be important for gaining and sustaining the jobs that will exist in the future, and retraining those workers who have been or are likely to be displaced from existing jobs.

Both these challenges will be addressed in subsequent chapters, but at this point it is useful to understand the changes that have been made in our education system over the last decade or two, particularly in the areas of mathematics and science. Unfortunately, some of these changes have not encouraged young Australians to acquire the knowledge and skills that will be of increasing importance in fields such as computation, artificial intelligence, robotics, nanotechnology and biotechnology. In fact, a recent report by the Australian Academy of Science indicates a decline in the percentage of students undertaking intermediate and advanced mathematics and an increase in the percentage taking only elementary mathematics.[14]

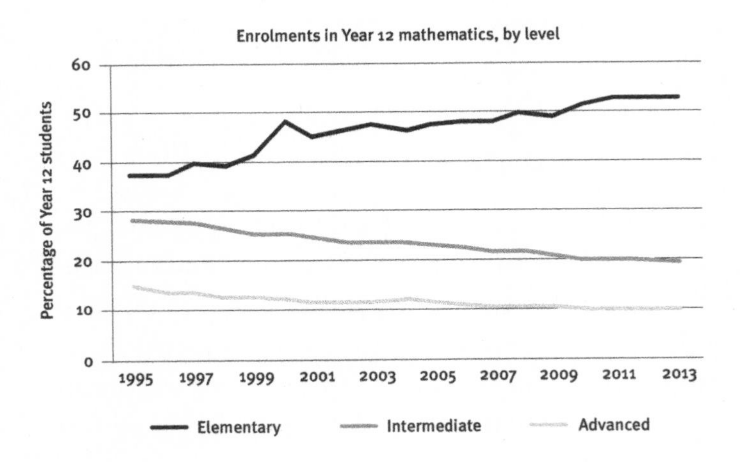

Source: Michael Evans and Frank Barrington, 'Year 12 Mathematics Participation Rates in Australia 1996–2015', Australian Mathematical Science Institute.

Not only do we see these rather discouraging trends over the last decade or two, but in some states the percentage of students undertaking no mathematics at all in their final years of high school has been increasing. In the 2001

NSW Higher School Certificate, 9.5 per cent of students did not take mathematics. By 2014, this percentage of students had more than doubled to 23.4 per cent.

Why does this matter, if only a small percentage of Australian students are going to end up as high technology researchers or in product design and development of advanced technology products?

It matters because mathematics underpins many of the physical, social and economic sciences on which our economy now depends. Mathematics is also essential for understanding many of the complex systems and technologies that surround us – including information and communication technologies, financial systems, energy and transport technology. Many of the important decisions that are made regarding health and other social service systems, as well as ecological and climate systems, depend on the application of statistics, which requires a solid foundation in mathematics. In fact, almost every area essential to our continuing prosperity is based on mathematics in some way.

It is for this reason that we will be focusing in later chapters on the importance of education and training, and the steps that could be taken to improve the preparedness of Australians to contribute to a future in which technology will play an increasing role.

We both believe education and training is a high-priority issue because we both believe that it is a moral imperative and also economically sound for policymakers to ensure that all Australians have the opportunity to participate in meaningful work.

Work is fundamental to self-worth and wellbeing, beyond the narrow and dry considerations of labour market participation. For most people, employment is more than just the main source of income and the basis of material prosperity. It also meets important social and psychological needs, including a sense of identity and social status. Meaningful work is important to people's happiness and prosperity. The evidence is clear that long-term unemployment negatively affects an individual's health.[15]

This raises the question of what constitutes meaningful and satisfying work – when machines equipped with sophisticated artificial intelligence are able to perform many of today's tasks, including some tasks undertaken by highly educated professionals, faster, with greater accuracy and at much lower costs than humans.

Chapter 2 covers the major technologies that are likely to impact on Australian society in the coming years – recognising both their potential benefits and adverse consequences. We will draw a comparison with what has happened in previous technology introductions and their effects on the employment environment as those technologies were widely deployed.

The third chapter builds on some of the research done overseas and in Australia to provide insight into how jobs in Australia are likely to evolve in a technology-dominated economy. We review a number of models that attempt to predict the percentage of jobs likely to be lost as a result of the growth in capability of AI and robotics technology.

In Chapter 4 we ask the question: 'Do the technological advances we see ahead of us have to lead to increased inequality, or can they instead lead to increased opportunities for all Australians?' We discuss the effects of both labour-saving and labour-linking technologies, and also the changing employee–employer power balance that is emerging.

Then follows a chapter on the skills and knowledge that need to be acquired to best prepare Australians for participating in a technology-based economy. The second part of Chapter 5 looks at the changes that have taken place in our education systems over the last two decades, and reflects on whether these changes have helped or hindered us in preparing for the evolution that is now occurring. We suggest seven ways our education system could be improved to prepare for the changes ahead.

In Chapter 6 we look at the challenges faced by federal and state governments as a result of the impact of digital technologies. We then address, through twenty suggested policy directions, actions governments could consider that would best position Australians for a fulfilling and meaningful life in a future in which our current material needs may be met with relatively fewer people in the workforce.

Chapter 7 suggests six actions that all of us, as individuals, can take to prepare for a more technologically oriented world.

CHAPTER 2
DISRUPTIVE TECHNOLOGIES AND LESSONS FROM THE PAST

Technologies have benefited humanity for much longer than there has been recorded history. Thousands of years before record-keeping began, Stone Age people were applying primitive technology to the domestication of plants. This 'First Agricultural Revolution', which took place about 12,000 years ago, allowed humans to move from nomadic hunter-gathering to the cultivation of food crops. More food production resulted in more people, and enabled a transition to living in settled societies. Social and political structures could develop and specialisation of jobs could take place.

The new stability and the division of labour meant some people had time to trial new ways of doing things. So, over the next several thousand years, knowledge advanced and new technologies came into existence, including the wheel, writing, mathematics, agriculture and bronze smelting.

The conventional view is that these technologies brought benefits to most people. And it is certainly true that the agricultural developments allowed many more people to be fed. But it is interesting to contemplate how the daily life of an average person in these agricultural communities compared with that of their hunter-gatherer predecessors. The Israeli popular historian Yuval Noah Harari argues that:

> Rather than heralding a new era of easy living, the Agricultural Revolution left farmers with lives generally more difficult and less satisfying than those of foragers. Hunter-gatherers spent their time in more stimulating and varied ways, and were less in danger of starvation and disease. The Agricultural Revolution certainly enlarged the sum total of food at the disposal of humankind, but the extra food did not translate into a better diet or more leisure. Rather it translated into population explosions and pampered elites. The average farmer worked harder than the average forager and got a worse diet in return.[1]

Over the grand sweep of time, however, there is little doubt that technology has been a boon for humankind. One only has to think about the rates of infant mortality, the suffering due to disease and illness and the general ignorance of the workings of the world that was the norm in centuries past. Innovation required the stable societies produced by that First Agricultural Revolution. But at the same time the introduction of major new technologies has caused disruption to societies and, in some cases, hardship in the short or even mid-term.

As we sit here in the beginning of the twenty-first century and try to forecast the implications of new technologies, it is useful to review the effects of previous major technology transitions on the wellbeing of societies.

We need to keep in mind, as French historian Fernand Braudel has observed, that before the nineteenth century, innovation and new ideas took a long time to penetrate societies.[2] For example, John Harrison's eighteenth-century invention of the 'compact chronometer', which was capable of keeping accurate time at sea, only became widely used many years after its invention. When Scottish naval doctor James Lind discovered in the mid-eighteenth century that citrus fruit could prevent scurvy, it took forty years before the Royal Navy began to regularly provide lemon juice to sailors. There were even times when innovations were lost, so technological progress was sometimes reversed. A very different situation exists today; new technologies are usually adopted very rapidly.

THE SCIENTIFIC REVOLUTION

Even though primitive technologies were used more than 10,000 years ago, the development and application of technology really took off following one of the major turning points in human history: the Scientific Revolution.

In the Scientific Revolution a different way of thinking emerged that no longer looked to ancient texts – especially religious texts – to explain the universe. There was a growing willingness to admit a lack of knowledge and understanding, to challenge traditional beliefs and concepts and to take an empirical approach, carrying out observations and conducting experiments. The outcome was the development of the 'scientific method', which was amusingly explained in a 1964 lecture by the famous American physicist and Nobel Prize winner Richard Feynman:

> In general, we look for a new law by the following process. First, we guess it [audience laughter], no, don't laugh, that's really true. Then we compute the consequences of the guess, to see what, if this law we guess is right, to see what it would imply. And then we compare the computation results to nature, or we, say, compare to experiment or experience, compare it directly with observations to see if it works.
>
> If it disagrees with experiment, it's wrong. In that simple statement is the key to science. It doesn't make any difference how beautiful your guess is, it doesn't matter how smart you are who made the guess, or what his name is ... If it disagrees with experiment, it's wrong. That's all there is to it.[3]

There are varying opinions as to when the Scientific Revolution began. There are even some historians who question whether there really was a major shift in thinking, rather than a gradual change. Most scientists, however, have little doubt that a revolution in the way knowledge was acquired took place in the sixteenth and seventeenth centuries.[4]

It is hard to overstate the long-term impact that the Scientific Revolution had on human progress. It heralded a change in the way intellectuals sought to discover the workings of the universe. Together with the use of mathematical models and tools to build theories from which predictions could be made and tested, this gave modern science the intellectual power to create and utilise new technologies.

BRITISH AGRICULTURAL REVOLUTION

But it was the British Agricultural Revolution, beginning in the mid-seventeenth century, and then the Industrial Revolution, which got underway in the eighteenth century, that had a direct impact on the life of the average person. The British Agricultural Revolution took place many millennia after the original Agricultural Revolution and had a startling impact. From the mid-seventeenth century, improvements in agricultural techniques, some influenced by the scientific method, allowed for a rapid increase in the British population – from a little over

5 million in 1700 to just under 8 million by 1800 and approximately 15 million by 1850. Driving this – and the rapid advance of the British economy – were improvements in tools such as the plough, the introduction of crop rotation techniques, selective breeding, improved transport, increasing private ownership of land, and laws allowing the fencing of common land into larger plots, thereby increasing efficiency.[5]

Like the original Agricultural Revolution, the British Agricultural Revolution was beneficial in the long run. But the lives of the large number of poorer agricultural labourers were not improved in the short to medium term. As the need for farm labour decreased, labourers had to look for other ways to feed their families. With more food being produced and a larger and better-fed working population, the scene was set for the Industrial Revolution.

INDUSTRIAL REVOLUTION

Beginning in about 1760, the Industrial Revolution saw hand-operated or manual methods progressively replaced by mechanised production. Greater volumes of machinery were produced. This drove agricultural productivity and, as a consequence, led to further falls in the number of agricultural workers.

The Industrial Revolution, even more than the British Agricultural Revolution, utilised science and technology to develop Britain's economy. Much of scientist John Smeaton's work, for example, found very practical application, such as his experiments in 1754 on the efficiency of waterwheels. The name most often associated with the Industrial Revolution is that of Scottish inventor James Watt, who, together with his collaborator Mathew Boulton, designed a successful steam engine. There were also significant improvements in textile manufacturing, coal mining and the production of chemicals, iron, steel, glass, paper and cement. While the Industrial Revolution was built on the scientific thinking that emerged from the Scientific Revolution, it was hard-nosed economics that drove the inventiveness of this period.

So what was the impact of the Industrial Revolution on the working life of the average person? It is generally accepted that living standards rose after 1850. But there is still debate about whether life for the average person got better or worse from 1760 to the 1840s. During this time there was slow growth in real income and consumption, and increasing income inequality. After 1830 these trends started to improve, although the wage increases might have come at the cost of an unhealthier and less safe work environment, longer hours and harder work.

Even though the Industrial Revolution did not result in increased unemployment, it brought significant social changes. Those employed in cottage industries and small workshops, particularly women, suffered most, as capital was invested in mechanisation and factories were established. The Ten Hours Act of 1847 limited the number of hours worked by women and children in the textiles industry to ten per day.

Economic historians Joel Mokyr, Chris Vickers and Nicolas Ziebarth have explained why the technological changes of the Industrial Revolution did not cause overall unemployment to rise, as some economists at the time predicted. Mechanisation in this period could only replace people for a limited number of tasks. Technological changes also spawned new products that required new industries and hence new jobs.[6]

Does the experience of the Industrial Revolution provide clues to what we can expect today and tomorrow with advances in robotics and AI? Mokyr and his co-authors caution that we cannot assume the pattern will repeat itself. 'It seems frighteningly plausible,' they argue, 'that this time will be different, and large sections of the labor market will be dislocated.' However, they are optimistic about the long-term benefits of technological advances while being realistic about the consequences for workers and the need for governments to take an active role:

> As has been true now for more than two centuries, technological advance will continue to improve the standard of living in many dramatic and unforeseeable ways. The law of comparative advantage strongly suggests that most workers will still have useful tasks to perform even in an economy where the

> capacities of robots and automation have increased considerably.
>
> The path of transition to this economy of the future may be disruptively painful for some workers and industries, as transitions tend to be. However, while the earliest transitions such as the Industrial Revolution were done with little governmental support for those displaced, this one will require public policy to ameliorate the harshest effects of dislocation.

Another interesting feature of the Industrial Revolution is that although Britain was the first to industrialise, by the second half of the nineteenth century it was beginning to lose its 'first mover' advantage and other countries were catching up. Between the Great Exhibition of Works of Industry of All Nations, held in the Crystal Palace in Hyde Park, London, in 1851, and the International Exposition (of Art and Industry), held in Paris in 1867, there was a visible decline in leadership of British technology. Belgium, Germany and the US were developing industries on a wide scale. Belgium had a similar advantage to Britain with its readily available coal. The US took the lead in mass production. Germany moved ahead in chemicals and the electrical industry. As we will see, this was due, at least in part at least, to the approach these countries took to education.

ELECTRIFICATION

While the Industrial Revolution transformed not just Britain but the entire world, it was not the last radical change. The next major step was the replacement of steam power by electrical power. This electrification of industry, railways and households had a dramatic effect on living standards, as both household lighting and household appliances became common. As in the Industrial Revolution, these changes also created new industries, products and jobs.

Major developments also took place in transport – road, rail, maritime and aviation – which had a transformative effect on industry, trade and people's lives, including new opportunities for employment.

Each of these very significant developments was dependent on technologies underpinned by science. This is particularly the case for the Digital Revolution, which is still underway today.

THE DIGITAL REVOLUTION

The Digital Revolution, which had its genesis in advances in communications and computation, is often also called the Computer Revolution or Information Revolution. Its beginning, like all technological revolutions, is hard to date precisely. But two seminal papers written about a decade apart were hugely influential.

The first, published in 1936, was written by the man often called the father of artificial intelligence, Alan

Turing.[7] He described a simple but universal machine capable, in principle, of undertaking the type of computational tasks done by today's powerful computers. Alonzo Church, Turing's colleague, called this device a 'Turing machine'. The importance of such an abstract and universal device is seen in the Physical Church–Turing Thesis, which says that 'any function that can be computed by a physical device can also be computed by a Turing machine, and more generally, that any physical system can be simulated by a Turing machine to any desired accuracy'.[8] This concept, revolutionary in its time, is vital to the tools of our modern world, with its heavy reliance on computation.[9]

Despite Turing's important work, he cannot be said to be the 'inventor' of the computer. That accolade goes to Charles Babbage, who in the middle of the nineteenth century designed an 'Analytical Engine' that had the important functional building blocks of a modern computer. Ada Lovelace, daughter of the English poet Lord Byron, also played an important role writing the first algorithm that could be programmed to run on the Analytical Engine.

The second paper that had a profound effect – this time in the field of electronic communications – was a 1948 publication by the American mathematician Claude Shannon.[10] In it, Shannon established that information could be defined and measured, and that for any communication channel there exists a maximum capacity to transmit information reliably. He thereby established the theoretical basis for digital communications and information theory. Even before this trailblazing work, Shannon had shown in his masters thesis how digital circuits could be analysed mathematically, which is essential for the design of the digital devices that power today's computing and communications systems.

While Turing and Shannon made fundamental contributions, the invention of the transistor in 1947 allowed modern computing and communications systems to be cheaper, smaller and more reliable. American inventors John Bardeen, Walter Brattain and William Shockley won a Nobel Prize in Physics for the work that led to the first transistor being built in Bell Labs in Murray Hill, New Jersey, just before Christmas 1947. Bardeen also won the 1972 Nobel Prize in Physics for his work on the theory of superconductors. This technology is used in Magnetic Resonance Imaging (MRI) machines in hospitals, in Maglev trains and in the Large Hadron Collider. Invented in 1960, the laser is another fundamental component of modern communications systems, and its invention, like that of the transistor, was dependent on a deep understanding of the physics of materials.[11]

From these beginnings, the advances in computation and communications have continued unabated for decades. A very significant evolution occurred in the 1980s, with the ethernet protocol being standardised in 1980 and the World Wide Web being invented in 1989.

COMPUTERISATION AND POLARISATION

So what have been the effects on employment and living standards from this revolution in communications and computation?

Analysing US data from 1920 to 2010, Lawrence Katz and Robert Margo show that:

> Changes in the organisation of work associated with computerization raise the demand for the cognitive and interpersonal skills used by highly educated professionals and managers and reduce the demand for the routine analytical (nonmanual) and mechanical (manual) skills that characterize many middle-educated, ordinary white-collar positions and manufacturing production jobs.[12]

They go on to observe that computerisation has had less direct impact on the demand for non-routine manual skills in low-wage jobs in the service and building sectors. They also note that overall wage inequality in the US has increased sharply since 1980. This supports the 'hollowing out' or 'polarisation' trend noted in our introduction. The picture in the US is that while the share of middle-skill, middle-income jobs has declined, there has been an overall growth in employment.

The same is true for the UK. A recent report by Deloitte shows employment in the UK grew by 23 per cent between 1992 and 2014, but there were large shifts in job types, due not only to computerisation and communications but also to other technologies and globalisation.[13] The report sounds a cautionary note:

> If the pace of adoption of technology is accelerating, society will need to prepare for higher levels of technological unemployment. And the way in which change increasingly rewards high-level education and skills suggests that income inequality may yet widen. Rapid advances in technology mean that education, training and the distribution of income are likely to be central to the political debate for many years to come.

A recent report written by Michael Coelli and Jeff Borland concluded that Australia has also experienced job polarisation. The following table shows the data supporting this effect in both Australia and Europe.[14]

Changes in employment shares by occupational skill level, Australia and Europe

	Lowest pay occupations	Middle pay occupations	Highest pay occupations
Australia 1966–2011	+2.2	-19.2	+17.0
Australia 1991–2011	+1.5	-8.5	+7.0
Europe average 1993–2010	+2.7	-9.9	+7.2

Source: Michael Coelli and Jeff Borland, 'Job Polarization and Earnings Inequality in Australia', 2016.

The researchers conclude that:

> Falling computer prices have caused rapid adoption of computer technology. The new technology has replaced routine cognitive and manual tasks previously undertaken by middle skill workers. At the same time, computer technology has been complementary to the non-routine cognitive and interactive tasks undertaken by high skill workers, thus raising their productivity and in turn the demand for these workers.

They also recognise that offshoring – basing a company's services overseas – was a contributor to this trend.

A Bridgewater Associates chart, published in 2016 and based on US and European data from MIT economist David Autor, represents quite dramatically the decline in middle-skill jobs in the 26-year period from 1990.[15]

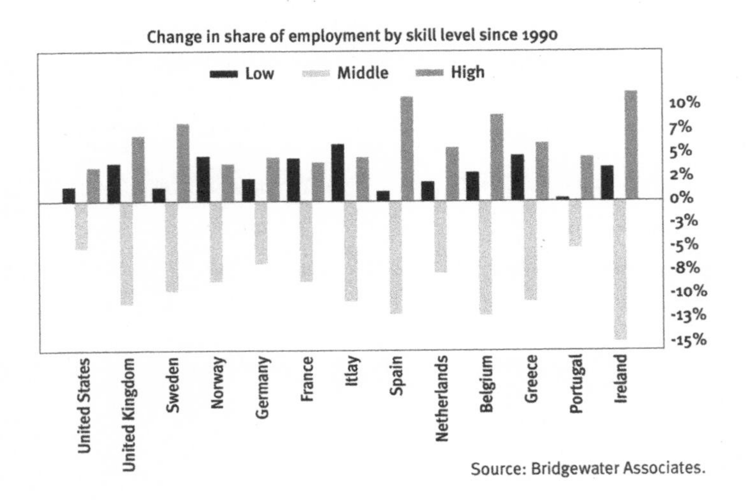

Source: Bridgewater Associates.

THE EMERGENCE OF ARTIFICIAL INTELLIGENCE

Science and technology are now interdependent, with one positively reinforcing the other. Scientific discovery now often relies on highly sophisticated technological tools, which are themselves masterpieces of engineering based on scientific principles. Consider the advances in particle physics research, for which the Large Hadron Collider has been pivotal, or the advances in genetic science enabled by machines that allow DNA sequencing to be done at a fraction of the time and cost of only a few years ago. Robotics, AI, virtual and augmented reality and cryptography rely on greater computational power. Emerging cellular and molecular biotechnologies, including gene therapies and bioinformatics, similarly harness the large computational resources and the ability to rapidly transfer huge files of biological data via modern fibre-optic and wireless communications.

Biological advances will have far-reaching effects on society and our health, longevity and quality of life – fascinating topics for another time. Our focus from here will be on artificial intelligence and robotics, which are likely to have the most profound effect on work and income.

AI has a relatively short and rather chequered history. Two pioneers were Warren McCulloch, an American neurophysiologist who also worked in the area of automated control systems, and Walter Pitts, an American logician who worked on computational neuroscience. In 1943 they jointly published a paper proposing a computational model for artificial neural networks. While the artificial neurons that formed the basis of their neural network were much simpler than real biological neurons, researchers nevertheless showed that such network architectures could carry out computations and learn. Subsequent work built upon these original ideas.

Around the same time, Turing, who in 1950 was giving lectures on AI, published his famous article 'Computing Machinery and Intelligence', where he laid out the ideas behind machine learning.

In 1956, a two-month AI workshop was held at Dartmouth College in the US. It was in the preparation for that workshop that the term 'artificial intelligence' was first used. This workshop is now seen as the founding event for AI. The ten people who attended and their students led the field for the next couple of decades.

Some of the early successes in AI research resulted in predictions being made that turned out to be wildly optimistic. The promise of success prompted US government investment to solve problems such as the translation of Russian into English during the space race. The difficulty of the problem was greatly underestimated and, by 1966, it became clear that the AI technology available at that time would not produce a useful translation tool.

The experience was similar in the UK, with the government dramatically reducing its funding for research when the difficulty of the problems AI was trying to solve became apparent. Most of the early programs worked well for relatively simple problems, but the methods could not be scaled up.

Following these setbacks, from around 1969 AI researchers had some success applying specific domain knowledge to problems. This heralded the era of 'expert systems', which did reduce costs for some administrative work and even had limited applications in medical diagnosis. These software systems generally used a knowledge base of special-purpose rules, which were usually provided by human experts. Prospector, for example, was an expert system developed in the late 1970s that evaluated the potential of a geological site to contain a valuable ore body. MYCIN was another early expert system, which diagnosed blood infections based on rules obtained by interviewing medical experts. It performed better than junior doctors.

As a result of these successes, money again flowed into AI research at the beginning of the 1980s. But by the end of the decade many of the claims regarding what the technology could achieve turned out, once again, to be much too ambitious. Thus began what is often referred to as the 'AI winter'. The algorithms were not sufficiently well developed, the computing power that was needed did not yet exist and the very large structured data sets on which machines could learn were not available.

But this was again only a temporary setback. Researchers started to incorporate developments in other fields of knowledge and to apply the rigours of the scientific method. Advances in statistics, information theory, optimisation and control theory and mathematical modelling were applied. Yoshua Bengio, a computer scientist at Montreal University, has observed in a recent *Scientific American* article that AI's outlook changed 'spectacularly' from 2005: 'That was when deep learning, an approach to building intelligent machines that drew inspiration from brain science, began to come into its own. In recent years deep learning has become a singular force propelling AI research forward.'[16]

'Deep learning' is the idea of artificial neural networks that had been worked on decades earlier. It allows AI programs to learn from the examples contained in the large data sets that are now available. This improved ability to learn has meant that researchers' ambitions in previous decades could begin to be realised, leading to advances in robotics, speech recognition and computer vision.

Bengio cites a particular type of artificial neural network, called a convolutional neural network, that has given the AI field a new impetus. These networks have many layers of artificial neurons, their architecture inspired by the part of the brain that processes the visual input from the eyes – the multilayered visual cortex. As a consequence, these networks are, for example, often able to recognise that different photographs of a face are the same person, even when the photographs have been taken from different perspectives. The 'deep' in 'deep learning' refers to the number of layers of artificial neurons; this layered structure allows these networks to do a better job of learning about their environment.

In contemplating what can be expected from AI technology in the years ahead, we need to be mindful of two issues. First, as we have seen, the field has been fraught with overly optimistic predictions in the past. Those working on AI will hopefully be careful not to make the same mistake again. Much has been written about the imminent arrival of driverless cars, for example, but the technical problems to be overcome to make this a practical reality are still significant. Designing cars in which the artificially intelligent system retains full control of the vehicle in all road, traffic and weather conditions and never relies on the backup of a human driver is a challenge, to say the least. It is a considerably more difficult and complex problem than designing an AI system to fly a plane. It is also still prohibitively expensive.

While fully autonomous vehicles may not be available for many years, it is likely we will see more use of driverless cars in restricted environments such as parking garages, or self-driving trucks that operate point to point on large highways where they do not need to deal with the multitude of conditions experienced in local environments.

Second, researchers may make a mistake in the other direction and not sufficiently prepare for dramatic increases in AI capability. It's possible it could ratchet up its own performance very quickly to an intelligence level that becomes difficult for humans to control.[17] This scenario is not the focus of this book and yet, because of its potentially harmful consequences, it is worth considering this issue – and why we should want our governments to pay attention to it.

FUNDAMENTALS OF ARTIFICIAL INTELLIGENCE

Central to AI is the concept of an 'agent', an entity that perceives, via sensors, the environment in which it exists and can take actions on the environment through actuators.

A web crawler that can browse the World Wide Web autonomously to gather data, a chatbot that can simulate a human conversation and a Twitterbot that can post autonomously onto Twitter are all examples of software agents. The sensory inputs and the actuators for a software agent are the usual input/output interfaces seen on a computer – keyboards, data transferred over a data link, scanned servers, displays, printers and data files.

Robots are also agents, but ones that carry out physical actions in the real world. They have sensors such as cameras, radar and lidar (similar to radar but using light from a laser rather than radio signals). They are often also equipped with accelerometers, GPS and gyroscopes to sense their own position and motion. Their actuators are usually called effectors, because they exert a physical force on their environment. These effectors include mechanical hands or grippers, legs and wheels. In the case of robotic apple pickers, the effectors include vacuum tubes to collect the apples; for robotic surgeons, needles and tweezers.

A common misconception is that AI and robotics technology needs to have advanced to a level where it can perform any cognitive task that a human can perform before it will have a significant effect on jobs. As we will see in the next chapter, this is not the case. We are still a long way from being able to fully emulate human intelligence. In fact, many researchers do not see the emulation of human intelligence as a primary driver of advances in the field. A program that is capable of the type of broad intelligence that humans exhibit is called Artificial General Intelligence (AGI) or 'strong AI'.

As the authors of the standard text on AI, Stuart Russell and Peter Norvig, observe, there is a difference between acting humanly and acting rationally.[18] It makes sense to focus on developing a 'rational agent' – one that acts to achieve the best outcome or, when there is uncertainty, the best expected outcome. For example, if a child was to run onto the road in front of a car, the driver must decide whether to swerve into oncoming traffic or other pedestrians to avoid the child. Having an AI make this sort of judgment when it is in control of a self-driving car raises many ethical issues. While we do not explore this deep and complex problem in this book, resolving this sort of issue may impact the timeframe for the introduction of fully autonomous vehicles.

In 1950 Turing, trying to answer the question 'Can machines think?', proposed a test, now called the Turing Test. It's a sort of imitation game, as the title of the Hollywood blockbuster based on Turing's life references, in which a machine passes the test if a human interrogator cannot tell whether the answers, in response to written questions, have come from a human or a machine.

An extension of this test was proposed by Stevan Harnad, a Hungarian cognitive scientist. He added a requirement for perceptual abilities and the ability to manipulate an object. The extended test is called the Total Turing Test.

Russell and Norvig note that a machine, if competent in the following six areas, would likely pass the Total Turing Test:

- Natural language processing, to be able to communicate in English.
- Knowledge representation, to store what is known and learned.
- Automated reasoning, to answer questions and draw conclusions from stored information.
- Machine learning, to adapt to new circumstances and detect and extrapolate from patterns.
- Computer vision, to perceive objects
- Robotics, to manipulate objects and allow movement.

They also make the point that, rather than focusing on emulating a human, concentrating on the six areas listed above is more likely to lead to productive outcomes:

> These six disciplines compose most of AI, and Turing deserves credit for designing a test that remains valid 60 years later. Yet AI researchers have devoted little effort to passing the Turing Test, believing that it is more important to study the underlying principles of intelligence than to duplicate an exemplar. The quest for 'artificial flight' succeeded when the Wright brothers and others stopped imitating birds and started using wind tunnels and learning about aerodynamics.

A machine will not have to pass the Total Turing Test in order to have a significant effect on employment. Specialised robots with subsets of these six capabilities will be able to carry out tasks that normally require a reasonable amount of human dexterity. A good example of a specialised machine equipped with AI targeted at a specific task is that of an apple-picking robot.

For an apple-picking robot to be successful when working in an orchard, it needs to be able to 'see' apples that are hidden behind leaves and other foliage, and also needs to be able to steer its picking effector to each apple and pick it without damaging the fruit. Such a robot has now been developed, and will no doubt be progressively improved, reduced in size and made faster and more versatile.[19]

This seems to be a big step forward in productivity for what is an arduous task.[20] Dan Steere, the CEO of the company that developed this particular robot, points out that 'while orchard yields have significantly improved over the last two decades, apple-picking labour productivity has not'.[21] The job, after all, involves a high level of manual dexterity and physical stamina, neither of which, because they are human characteristics, is going to increase year after year (until or unless we develop the ability to enhance the human genome using gene editing techniques).

But it also begs the question: what other employment would be available to the workers who seasonally pick apples? An estimated 70,000 people pick apples each autumn in the US, and that is just 5 per cent of the worldwide apple production.

We can be reasonably sure that the capabilities of AI and robotics will continue to grow. Whether these capabilities replace existing jobs or are used to make existing jobs more productive and interesting – or, indeed, lead to the creation of new jobs – is the subject of the next chapter.

CHAPTER 3
HOW WILL EMPLOYMENT EVOLVE IN A TECHNOLOGY-DOMINATED ECONOMY?

Jobs will change in the coming years under the influence of new technology, especially AI and robotics. This raises five critical questions for Australians.

First, which jobs will be replaced by technology? Second, what new jobs are likely to be created? Third, if new jobs are created, are these likely to offset the number of jobs that will disappear? Fourth, over what time period is all of this likely to happen? And finally, what skills will be required to gain one of these new jobs?

Unfortunately, there is no general consensus on the answers, either for us here in Australia or for those in other countries facing the same somewhat uncertain future. Part of the uncertainty is the difficulty of predicting when some of the new technologies will be introduced. Consider the following.

Two large car companies are fairly optimistic with their projections about driverless vehicles. In 2015 the CEO of

Nissan said that 'after introducing autonomous technology and artificial intelligence one piece at a time across its line-up over the next few years, Nissan expects to offer vehicles capable of fully autonomous city driving by 2020'.[1] The BMW Group plans to 'develop the necessary solutions and innovative systems for highly and fully automated driving to bring these technologies into series production by 2021'.[2]

The Boston Consulting Group, however, predicted in 2015 that by 2035 only 10 per cent of new vehicles would have fully autonomous driving capability.[3] A 2016 *Scientific American* article by research engineer Steven Shladover is even more pessimistic: 'Fully automated vehicles capable of driving in every situation will not be here until 2075. Could it happen sooner than that? Certainly. But not by much.'[4]

While we cannot say who is right in this debate, we do know that the introduction of autonomous vehicles on Australian roads will adversely affect the large number of bus, taxi and commercial vehicle drivers currently employed. Of course, there are some whose jobs are already being disrupted by technology. Uber, a platform that aligns drivers and riders using a pricing and dispatch mechanism, is a good example. Uber's long-term vision is to replace human-driven cars with driverless vehicles.

WHICH JOBS?

The uncertainty around fully automated vehicles is an example of the wider debate regarding technology's impact on jobs. Numerous studies have been done on this subject in the last few years and their conclusions vary significantly.

As we've already noted, Carl Frey and Michael Osborne's 2013 paper was the most influential.[5] The pair used data from the US to estimate the probability of occupations being replaced by computerisation within a decade or two. They did not try to be precise on the timing, or to estimate the probability of new jobs being created. They saw their study as a guide to the number of jobs that would need to be created in the next few decades to offset those replaced by technology.

Frey and Osborne took account of the advances that have been taking place in the various branches of AI, particularly those focused on machine learning. These advances involve algorithms that allow cognitive tasks to be carried out by machines, and thus allow tasks normally done by humans to be automated. Machine learning is largely dependent, with today's technology, on having large data sets on which the learning algorithms can be trained – hence the importance of what is known as 'big data'.

It is important to point out that the shift to probabilistic or statistical computation, which is now driving advances in machine learning, is a different process to what most people understand when they think about the rule-based deductive logic that is central to many existing computer programs.

Machine learning makes use of inductive rules where the processing done by the computational engine cannot be specified by a sequence of pre-defined steps. It is more a process of pattern recognition: the machine gradually learns the associations between the input information the machine has available to it and the desired outputs. Judea Pearl, a professor of Computer Science at the University of California, Los Angeles, describes this process as an 'Engine for Evidence'.[6] We are seeing increasing success in areas such as machine vision, handwriting recognition and natural language processing. This is, in part, due to the incremental improvement of probabilistic algorithms, making use of new evidence or data. The massive data sets that are now available allow machines to be trained on millions of data points.

Frey and Osborne's models take into account three 'bottlenecks' in machine performance – those requiring perception and manipulation, social intelligence or creative intelligence. They believe that machines are not yet close to being able to match human perception, particularly in a very cluttered field of view as is experienced in an unstructured work environment. And while there has been considerable progress in the design of manipulators, there is nothing yet that comes close to the versatility and manipulative capability of the human hand.

Frey and Osborne's second bottleneck to computerisation – social intelligence – probably presents the most difficulty for AI. The recognition in real time of human emotions, including the subtle signals humans transmit via their body language, remains a challenge. Producing appropriate (human-like) responses is even more difficult.

Third, Frey and Osborne assumed that it is unlikely that occupations requiring a high degree of creative intelligence will be automated in the next few decades. This is probably reasonable, although the boundaries of what machines can achieve continue to be pushed.

In his recent book, Yuval Noah Harari cites the example of David Cope, a musicology professor at the University of California, Santa Cruz, who developed a program called EMI (Experiments in Musical Intelligence) that can compose classical music.

Another professor, Steve Larson from the University of Oregon, challenged Cope to a 'musical showdown'. He suggested that pianists play a piece by Bach, a piece by EMI and a piece by Larson himself. The audience would then vote on who composed each piece. Harari recounts the outcome:

> Larson was convinced people would easily tell the difference between soulful human composition, and the lifeless artifact of a machine. Cope accepted the challenge. On the appointed date, hundreds of lecturers, students and music fans assembled in the University of Oregon's concert hall. At the end of the performance a vote was taken. The result? The audience thought that EMI's piece was genuine Bach,

> that Bach's piece was composed by Larson and that Larson's piece was produced by a computer.[7]

So we may be surprised at how quickly artificial intelligence will be able to make inroads into what we would normally regard as creative tasks.

The model Frey and Osborne developed assumes that the likelihood of a job being automated increases as technology progressively overcomes these three bottlenecks. They also note that, as in the past, while machines may not be able to undertake a particular job fully, the task may be restructured with the aim of maximising the extent of automation that is possible.

They then used data from the US Department of Labor and, after ranking and aggregating, ended up with a data set of 702 occupations ranked on the basis of skills, abilities and knowledge.

Next, with the help of AI experts, they selected the seventy jobs whose risk of being automated they could be most confident about, then subjectively labelled them as such. The researchers relied on nine different variables – such as manual dexterity, persuasion and originality – stemming from the three bottlenecks to make their determination. Taking a smaller sample of occupations for which they were most confident reduced the risk of subjective bias affecting their analysis.

Frey and Osborne then used the data from these seventy occupations as test data in the development of an algorithm to determine how likely any particular job was to become automated.

The pair categorised the probability of an occupation being replaced by automation in the next few decades as low (less than 30 per cent), medium (between 30 and 70 per cent) or high (above 70 per cent).

The output of their model indicates that 47 per cent of jobs in the US are in the high-risk category, with at least a 70 per cent chance of being automatable within the next few decades. At the other end of the spectrum, some 33 per cent of US jobs are in the low-risk category, with a less than 30 per cent likelihood of being replaced by automation.

They used the same data to examine the relationship between the probability of a job being automated or computerised and two important characteristics of that job. The first is the median wage and the second is the educational level required for the job, measured by the proportion of job holders who have at least a bachelor's degree.

What Frey and Osborne found is that both salary and educational level are inversely related to the susceptibility to automation or computerisation. They predicted that computerisation will have the greatest impact, in the near future, on low-skill and low-wage jobs. This not only means fewer jobs for lower-skilled workers; it could also have flow-on effects for those squeezed out of middle-skill and middle-income jobs and left to compete with low-skilled workers for a diminishing supply of lower-paid roles.

In 2014 Frey and Osborne, in cooperation with Deloitte, analysed the risks to jobs in the UK due to automation, with a particular focus on London.[8] They used the same methodology, but employed the International Standard Classification of Occupations rather than US Department of Labor data. They reached similar conclusions to the US study, with 35 per cent of jobs in the UK being predicted to disappear in the next two decades as a result of automation.[9] The jobs most at risk in the UK were in administrative support, sales and service, transportation, construction and extraction, and manufacturing.

As was found using the US data, there was an inverse relationship between the jobs most at risk from automation and the salaries of those jobs. Jobs with salaries of less than £30,000 a year were almost five times more likely to be lost to automation than jobs with salaries of more than £100,000 a year.

In London, due to the large proportion of finance jobs and jobs in creative sectors and a lower proportion of manufacturing jobs, there is a reduced likelihood of job losses, with 30 per cent of jobs predicted to be at risk from technology over the next two decades.

In a recent report on Australia's future workforce, University of Sydney Professor Hugh Durrant-Whyte, now chief scientific advisor at the UK Ministry of Defence, together with his colleagues used the same methodology and initial data as Frey and Osborne.[10] They migrated the US job codes to those based on the

Australian and New Zealand Standard Classification of Occupations.

Durrant-Whyte and his team arrived at a similar conclusion for Australia. Forty per cent of current Australian jobs have a probability of greater than 70 per cent of being computerised or automated within the next ten to fifteen years. According to their findings, managers and professionals are least likely to have their jobs automated or replaced by computers, whereas labourers, machine operators and drivers, clerical and administrative workers, and many technicians and trade workers are highly likely to have their jobs replaced.

Durrant-Whyte and his colleagues took an additional step and looked at how the affected jobs are distributed across Australia. Areas with a high dependence on mining, such as Western Australia and Queensland, have a higher probability of job replacement by automation and computerisation. The CBDs and inner-city areas of the larger Australian cities will be much less susceptible, due to the much larger percentage of professional, technical and creative jobs in those areas.

TASKS WITHIN JOBS

It is important to understand the assumptions underpinning Frey and Osborne's model. For instance, they chose not to include several factors, such as the trade-off between labour and capital costs; the impact of regulatory or legislative processes; whether the public would accept particular technologies; and the possibility that a worker can go on to perform other important tasks if their current task were computerised. Each of these factors could have a significant influence.

Melanie Arntz, Terry Gregory and Ulrich Zierahn for the OECD argue that Frey and Osborne may be both overestimating the percentage of jobs lost and underestimating the percentage of jobs whose scope has changed:

> These studies ... assume that whole occupations rather than single job-tasks are automated by technology. As we argue, this might lead to an overestimation of job automatability, as occupations labelled as high-risk occupations often still contain a substantial share of tasks that are hard to automate.[11]

Their analysis relies on data from the International Assessment of Adult Competencies, which provides a comprehensive list of tasks that people perform on the job – reported by the people doing the job. This takes account of the fact that individuals within the same occupation often perform different tasks.

To distinguish between these two methodologies, the OECD authors describe their approach as 'task based' and describe the Frey and Osborne approach as 'occupation based'.

Arntz and her colleagues look at twenty-one OECD countries and conclude the average percentage of jobs at risk of automation across the OECD is 9 per cent and, in particular, 9 per cent for the US – compared with 47 per cent in the Frey and Osborne study. One of the main reasons Arntz and her colleagues give for this difference is that even for the high-risk jobs there are tasks, such as person-to-person communications, that cannot easily be done by machines.

They note that their low estimate may in fact be *overstating* the number of job losses, for a number of reasons. First, the estimates in all of the studies depend on technology experts predicting the progress that AI and robotics technology will make in the coming decades, and technologists have, in the past, tended to be optimistic in their predictions. Second, organisations and workers may adapt more quickly than anticipated to use the new technological capability as a complement to their job, rather than replacing it. Third, and as acknowledged by Frey and Osborne, the models used made no attempt to estimate the new jobs that may be created by the technologies.

But in both the Frey and Osborne and OECD studies it is the less-educated workers who will bear the brunt of technological change. The authors of the OECD study note that there will be a significant amount of up-skilling and retraining required for today's less-educated workers to be able to participate in the job market of the future.

Analysis published by the McKinsey Global Institute (MGI) in January 2017 took a similar approach to that used by the OECD study. MGI report that, globally, while almost half of the individual activities that make up jobs could be automated using today's technology, less than 5 per cent of today's jobs could be *fully* automated. But that's not to say they can't be partially automated; at least 30 per cent of tasks are automatable in more than 60 per cent of jobs.[12]

They also make the point that increasing levels of automation will increase productivity, which will be sorely needed due to the ageing populations in many countries.

The McKinsey analysts say the changes will play out over decades. They make the obvious point that if automation is to increase productivity, displaced workers will need to find alternative employment.

A recent comprehensive analysis by AlphaBeta also addresses the need to redeploy workers who will inevitably be displaced.[13] The strategy advisory company's *Automation Advantage* report, commissioned by Google, then goes a step further to predict that all jobs will change, not just a select few.

It's worth emphasising that AlphaBeta's findings are full of optimism, rather than anxiety, about the changes already underway. Although the report concedes some jobs will be lost to automation, it insists more will be created as technology provides more opportunities and redefines what is meant by 'work'. This positive outlook springs from the notion that automation won't just change what jobs Australians do, but how they do them. The consultancy firm considered the 20 billion hours Australians worked last year and broke that time down into 2000 specific tasks. Each of those activities was put into six categories: information analysis, predictable physical, unpredictable physical, interpersonal, creative and information synthesis.

The report found the relative importance of the different activity categories over the past fifteen years were predicted to continue for at least the next fifteen. The categories of interpersonal, creative and information synthesis were projected to increase from just under half of all work activity to almost 70 per cent over the thirty years from 2000 to 2030. A corresponding reduction in the other three categories – information analysis, predictable physical and unpredictable physical – was expected too.

AlphaBeta's analysis also suggested that, by 2030, machines would likely relieve workers of two hours' worth of their most repetitive manual tasks each week. That would allow workers to rely more on brains and personality rather than physical labour, and spend more time on activities that create the most value – those that fall in the interpersonal, creative and information synthesis categories. This shift from physical to mental tasks is expected to boost both job safety and job satisfaction.

Like us, AlphaBeta is convinced that we can only help those who will be displaced and take advantage of the new machine age if the right, forward-looking policy settings are in place.

In 2016 the Australian Federal Department of Employment projected growth in jobs until the end of 2020, following an increase of more than 700,000 jobs created between 2010 and 2015. Significant growth is predicted in four industries: education and training; healthcare and social assistance; professional, scientific and technical services; and retail trade; partly offset by much smaller declines in agriculture, forestry and fishing, and mining and manufacturing.[14]

WHICH INDUSTRIES?

Although the Australian economy and jobs are now dominated by the services sector, Australia is no less likely to be disrupted by automation or computerisation than other advanced Western economies. In fact, Australia's services sector is significantly smaller as a percentage of the economy than those of the US or the UK, and on par with Canada, Germany and New Zealand.[15]

In the past, job losses due to automation have largely taken place in the industrial sector and agriculture. But as a recent CSIRO report noted: 'The information revolution will not be limited to manual jobs. Its impacts will lie heavily, if not mostly, within the service sector industries that account for over two-thirds of the Australian economy.'[16]

Even jobs that one may think of as immune are at risk. Richard Susskind is the president of the UK Society for Computers and Law and holds professorships at three universities, including Oxford. He and his son Daniel Susskind, an Oxford economics lecturer, argue that professional jobs in areas such as law, healthcare, consulting and accounting are unlikely to continue in their current state. Major changes to these professions, they say, may in fact be beneficial to society: 'By and large, our professions are unaffordable, under-exploiting technology, disempowering, ethically challengeable, underperforming, and inscrutable.'[17] The challenge for the professions is to find a model that provides a reasonable income while evolving to a system that provides greater access to practical expertise at lower costs.

Richard and Daniel Susskind have little doubt that the work of professionals is going to be challenged by emerging technologies:

> We foresee that, in the end, the traditional professions will be dismantled, leaving most (but not all) professions to be replaced by less expert people and high-performing systems. We expect new roles will arise, but we are unsure how long they will last, because they too, in due course, may be taken on by machines.

These words reflect a critical difference between the technological revolutions of the past – the agricultural, industrial and digital revolutions – and the technological revolution we are now entering. Previously, increasingly sophisticated machines and equipment replaced many jobs that were dangerous, laborious or tedious. When new jobs were created they were often more interesting and less manually intensive than the jobs that were eliminated, and could be mastered by an increasingly educated population. Whereas today's sophisticated and complex jobs could give way to those supported by highly intelligent systems.

MAKING SENSE OF IT ALL

What should we make of these conflicting predictions? On one hand, we have several Frey and Osborne–style studies forecasting significant job losses. On the other, we have an OECD study, a McKinsey report and an AlphaBeta analysis predicting that while technology will have a significant effect on the types of tasks that are performed, this is more likely to result in scope changes to jobs than in jobs being lost. And in the short term we have a fairly optimistic view from Australia's Department of Employment.

We cannot know whether it will be 10 per cent or 40 per cent of jobs in Australia lost to automation and computerisation over the next two decades. And even if it is at the higher end, the new technologies could lead to many new jobs – as has happened in the past with other major technology introductions. It is also not yet clear whether the changes will result in a continuing hollowing-out of the mid-skill and mid-income jobs, whether the greatest impact will be on the low-skill, low-paid jobs, or even whether some high-skilled jobs previously thought to be immune will end up on the chopping block.

This uncertainty is evident also in the Pew Research Center report we cited earlier, where almost 1900 experts responded to a question that asked whether by 2025 AI and robotics will have displaced more jobs than they have created.[18] Slightly less than half (48 per cent) of the respondents foresaw a future in which significant numbers of both white- and blue-collar workers had been displaced by technology, leading to income inequality, large numbers of unemployable people and a breakdown in the social order. Slightly more than half (52 per cent) foresaw technology being able to do many of the jobs currently done by humans. But they believed that human ingenuity would create new jobs, new industries and new ways to produce an income, just as has happened with previous technological revolutions.

It is an unfortunate reality that we cannot say with any confidence which of these predictions is likely to eventuate. The differences of opinion are unlikely to be resolved until the data itself reveals the answer. That does not mean, however, that we need to sit on our hands and wait to see what happens. No doubt the new technologies will open up areas of opportunity for those who have the capabilities to develop and use these new technologies. But what of our fellow Australian workers who do not currently have these capabilities? The authors of all of these studies see the need for government policies that prepare for the future, with increased education, training and up-skilling as well as transition assistance for those displaced.

The Pew report elegantly argues that action could and should be taken by those in a position to do so:

> Although technological advancement often seems to take on a mind of its own, humans are in control of the political, social, and economic systems that will ultimately determine whether the coming wave of technological change has a positive or negative impact on jobs and employment.[19]

While the long time frame associated with this transition is good news, the weight of opinion is that this transition could further accentuate inequality in the Australian economy. When the effects of automation and AI impact unevenly on Australians who lack the resources or expertise to prepare or respond effectively, there is a clear role for Australian governments to play.

We began this chapter with five questions and so far we've focused on the current thinking on the first four. The fifth question, regarding the skills required to be successful in the future, deserves a chapter of its own. But first we turn to the critical question of how the changes to employment reviewed in this chapter are likely to disproportionately affect different sections of our community.

CHAPTER 4
THE FUTURE – MORE OPPORTUNITIES OR MORE INEQUALITY?

The costs and benefits of rapid technological change fall disproportionately on different people within our economy and our society. Not everybody cares about this, but we should. Few among us would want to see Tyler Cowen's vision realised of workers divided into two camps: one capable of working with machines, the other replaced by them.[1] Yet we are already seeing machine learning, robotics and artificial intelligence affect low-paid and low-skilled jobs. Skilled middle-income jobs are also at risk, and even some top-tier professions may not be immune. But the gains flow increasingly to the top. This is one reason increasing inequality stalks the developed world, but not the only one.

We care about technology's contribution to this problem because inequality is a threat to social wellbeing, to a growing economy and to intergenerational mobility, which is so important to dynamic, flourishing nations.

Increased inequality risks turning Australia into a country where a tiny minority gallops ahead while more and more people are left behind.

Economic inequality and immobility divides and diminishes our society and pushes people to the political fringes. The rise of populism, here and overseas, is fed by a dominant grievance: that the link between hard work and reward has been severed because the rules of the economy are written to benefit someone else. The rise of newer technologies has the potential to make this worse, creating clusters of people who lose out from change and see no role for themselves in a modern economy.

We believe that there is no such thing as 'technological trickle-down': concentrating economic power into fewer human hands will not deliver the security and prosperity our society needs. The changes we discuss in this book are not all negative, nor is their impact on equality and mobility necessarily inevitable or permanent. But we need to understand the dangers, and then care enough to act.

INEQUALITY AND IMMOBILITY

It is a remarkable achievement that Australia has maintained a record quarter-century of uninterrupted economic growth, especially when that period contained the biggest global downturn since the Great Depression. But the gains of this growth are increasingly flowing to the top end of town.

A recent Oxfam report estimated that, globally, just eight men own the same amount of wealth as the poorest half of the world's population, or 3.6 billion people.[2] The OECD warned in 2015 that 'inequality is bad and getting worse' within member countries.[3] It found:

- The richest 10 per cent of people in OECD countries earn nearly ten times more than the poorest 10 per cent (compared with seven times more in the 1980s).
- The richest 10 per cent in 2012 controlled half of all total household wealth, and the wealthiest 1 per cent held 18 per cent.
- The poorest 40 per cent only held 3 per cent of household wealth.

In the US, more than 50 per cent of total income went to the top 10 percent of income earners in 2012. The top 1 per cent earned 22 per cent of total income. What's more, between 1983 and 2009 the total value of US assets increased but those in the bottom 80 per cent saw their wealth decrease. This means that the top 20 per cent of individuals received more than the total increase in wealth over this period because the wealth of the bottom 80 per cent decreased.[4]

We take limited comfort from the fact that Australia's situation is not quite as extreme as that in the United States. But the trend is the same. The two richest Australians,

Blair Parry-Okeden and Gina Rinehart, own more than the combined wealth of the country's poorest 20 per cent. All up, the top 1 per cent is wealthier than the bottom 70 per cent of Australians.[5]

In fact, Australian Bureau of Statistics data on income inequality, which stretches back to 1975, shows real wages have grown for the top tenth of earners more than three times faster than for the bottom tenth. If the wages of the country's lowest earners had increased at the same pace as the highest earners', they would be $16,000 a year better off. In terms of wealth inequality, which measures the distribution of assets in the economy, the top 1 per cent of Australians have seen their share of wealth at least double since the late 1970s. At the very top, the richest 0.001 per cent of Australians have seen their share of national wealth triple since the mid-1980s.[6]

Of course, Australia has historically been insulated by a strong social safety net. This includes means-tested welfare, pensions and family payments, and a decent social wage, including high public investment in education and universal healthcare. Our industrial agreements, including solid minimum conditions, penalty rates and employment standards, have also somewhat buttressed us against rising income inequality. But the portents for the future are not good.

The idea that massive gains accumulated by the few at the top will trickle down to the rest of society has been long discredited. And it is glaringly obvious why inequality is bad for an individual's wellbeing – less pay, poorer health and lower living standards. What is not as uniformly agreed upon, though it should be, is the effect of inequality on the economy as a whole.

Perhaps the most troubling impact of inequality is how it cascades through families and neighbourhoods, sentencing one generation to the same, or a similar, economic fate as the one before.

Inequality in one generation breeds inequality in the next. Consider the 'Great Gatsby Curve', which charts the likelihood that a son will inherit his father's economic position in adulthood.[7] The curve shows that less equal countries tend to be those in which a greater fraction of economic advantage and disadvantage is passed on.[8]

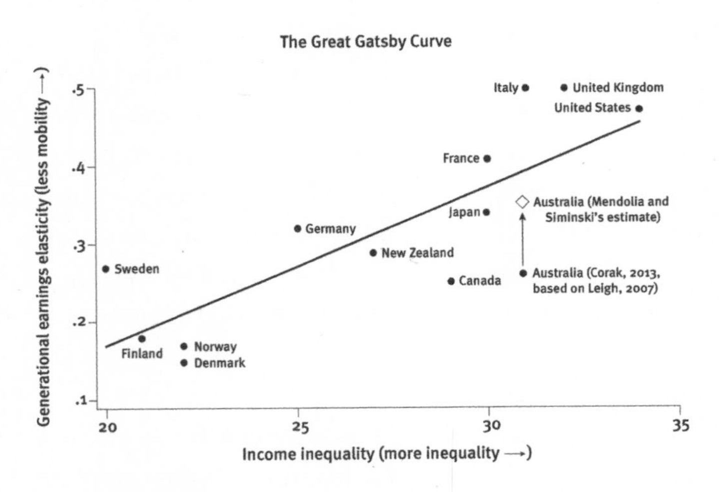

Source: Sylvia Mendolia and Peter Siminski, 'New Estimates of Intergenerational Mobility in Australia', 2016.

While Miles Corak's original graph shows intergenerational wages elasticity in Australia (around 0.25) is about half that of the United States (around 0.5), researchers Sylvia Mendolia and Peter Siminski have used more recent data to suggest an even bleaker reality. They estimated intergenerational wages elasticity in Australia to actually sit around 0.35. So an Australian parent earning 10 per cent less than average will have a child earning about 3.5 per cent less than average.[9]

Not only is a lack of social mobility unfair, it is also likely to make our economy less prosperous.[10] The OECD has found that countries that are becoming more equal grow faster than those where inequality is rising.[11] A nation comprised of people who have their prospects of success limited at birth will not generate the dynamism and creativity necessary to drive its economy. Where intergenerational mobility is low, a bright young person from a disadvantaged background will face career-limiting barriers. The skills and human capital of that bright but poor young person are lost to the economy altogether.

No society is better off if tomorrow's stock of doctors, engineers and entrepreneurs comes exclusively from the children of today's doctors, engineers and entrepreneurs.

LABOUR-SAVING AND LABOUR-LINKING

There is no argument that technology – whether it be through improvements in communications such as the smartphone, or medical advances – has made people's lives better in a general sense, both in Australia and around the world. Technology has actually reduced inequality between countries by opening new markets and bringing opportunities to poorer countries. According to the International Monetary Fund, new technology has substantially improved productivity and collective well-being. But it has also been a key driver in widening the inequality gap. In fact, the IMF argues that 'technological advances have been found to have contributed the most to rising income inequality in OECD countries'.[12]

As Andrew Leigh points out, it's not enough to think about the aggregate benefits when the benefits are distributed unequally.[13] The pride we take in being an egalitarian nation, a country that values the 'fair go', means we cannot just revel in the broader benefits while ignoring the impacts on our most vulnerable – those either left behind or hindered by technology's march.

That means understanding the links between technological advances and the risks of further inequality and immobility.

In 2016 the World Bank's chief economist, Kaushik Basu, wrote of a 'persistent recession' – the ongoing global economic slowdown sparked by the financial crisis of 2008.[14] Basu attributed the slowdown after the GFC to a deeper shift in the global economy driven by two technological innovations. These innovations are 'labour-saving' – the use of inventions that reduce the time or effort taken to complete a task, such as industrial robots – and 'labour-linking', which allows people, often via the internet, to work for employers and firms in different countries without needing to migrate.

Labour-saving and labour-linking disproportionately threaten the livelihoods of key groups – such as unskilled workers, the middle classes of advanced economies and older men in traditional occupations – in at least four ways.

First, technology changes the composition of the workforce. In the simplest version, it eliminates jobs, directly leading to unemployment and lower wages. But it is not as clear-cut as that. Economic historian James Bessen[15] and others argue automation is not actually causing a net loss of jobs, but it is disproportionately displacing workers in low-wage occupations, who are much less likely to use computers than their high-wage counterparts. In effect, some jobs are eliminated, pushing middle-income earners to compete for an ever-dwindling supply of lower-income roles and squeezing low-income workers from those roles. Bessen says that 'the net effect implies a substantial dislocation of work to higher-wage occupations'.

This leads to employment polarisation in high-income countries such as Australia, which the World Bank has linked to declining labour shares in national income.[16] In its *World Development Report 2016*, the World Bank notes that employment was growing in high-skilled and high-paying occupations, like managers and professionals, and also in low-paying and low-skilled jobs, like service and sales workers. But workers such as clerks and machine operators – those in middle-skilled and middle-paying roles – were being squeezed out.

This hollowing-out doesn't mean the middle class are necessarily ending up without work. It means they have 'cascaded down to compete with those in low-paid jobs'.[17] The end result is an hourglass economy made up of high-income and low-income earners, with fewer in the middle.

Second, technology changes the location of jobs. Emerging economies such as China, India and Malaysia are already using labour-linking technology and can supply lower-cost labour. The same is not true of developed nations such as Australia, which are shedding jobs offshore.

The effect of labour-linking technology is most clearly seen in global value chains, which include all the elements involved in the production of goods or services where different stages are located across different countries. Labour-linking has meant the relative rewards for high-skill functions like research have increased, and those for lower-skill or manufacturing jobs have decreased over time.

We see this dynamic in the 'smile curve' of global value chains, which graph the value added at various points along the chain, such as research and development; manufacturing; and branding and marketing. With many manufacturing jobs being outsourced or heading offshore, some economists argue that the curve has deepened among OECD countries, meaning that instead of value being evenly spread along the chain, it is now more U-shaped, with fabrication and assembly accounting for a much lower share of value added.[18]

Third, technology changes the skills required to succeed. Harvard economists Claudia Goldin and Larry Katz make the case that inequality is determined by a race between technology and education.[19] Technological change, they argue, creates 'winners and losers', particularly when new technologies increase the demand for higher-skilled workers. This skills premium risks creating a divide where the income of a few increases far more quickly than that of others. But they do not consider this divide to be inevitable:

> If workers have flexible skills and if the educational infrastructure expands sufficiently, then the supply of skills will increase as demand increases for them. Growth and the premium to skill will be balanced and the race between technology and education will not be won by either side and prosperity will be widely shared.[20]

Unfortunately, that symbiosis is a long way off. The increased premium modern technology places on skills fuels inequality.[21] That is because it substitutes for many unskilled workers' jobs and tasks, whereas it acts as a complement for some skilled workers. Also, unskilled workers may not be able to afford to upskill or retrain.

Finally, technology changes the power relationship between worker and corporation. Digital advances are likely to see 'non-standard' forms of work and shorter job tenures become more common, particularly among the young.[22] The rise of the 'gig economy' indicates this is already occurring, but is only one, more recent, part of a much broader story about contracting, casualisation, temping and the like.[23]

Internet-enabled jobs can provide workers with more flexibility and even generate new opportunities for the unskilled or those who want to vary the amount of work they do.[24] But unions and the World Bank have cautioned that these kinds of jobs can also erode workers' rights by reducing their bargaining power and not providing sufficient benefits.[25]

All of these changes – workforce composition, where jobs are located, the skills premium and the shifting power relationship – feed into a pressing concern for individual workers: wages. The way these changes flow through to lower wages is, mostly, self-evident: by pushing middle-income workers down to compete for lower-skilled jobs, or by increasing lower-cost labour from emerging nations, or by eroding workers' bargaining rights. But there are less obvious impacts too. For instance, Bessen argues greater inequality of wages can arise within occupations if new skills are costly or difficult to acquire.[26]

Of course, there is also a significant impact on wages at the top end of the scale.[27] Even a street poll asking people to name well-known billionaires would quickly demonstrate the link between technological innovation and extremely high wages, with the likes of Bill Gates, Mark Zuckerberg and Elon Musk dominating public perceptions.

There is nothing wrong with creating wealth through innovation; it should be actively encouraged. And the rewards that come with it should flow through to innovators and inventors, as has happened in the past. But wealth is even more skewed towards the top with new tech companies, either because relatively few jobs are created or because many low-paid casual jobs are. In 2013 Facebook's workforce was fewer than 8000 people, compared to IBM with approximately 430,000 and Dell with about 109,000. When Amazon bought video-streaming company Twitch in 2014 for a little under US$1 billion, it had 170 employees. When Instagram was bought by Facebook for US$1 billion in 2012, it employed only thirteen people.[28] It is true, however, that companies such as Instagram have created work for non-employees – artists and photographers, for example.

These technological changes are partly to blame for workers' wages making up a significantly smaller slice of the economy. According to Basu, it's a trend occurring globally, 'at rates rarely witnessed'.[29]

A 2017 report by The Australia Institute's Centre for Future Work found that labour compensation as a percentage of GDP hit 46.2 per cent in the March quarter – the lowest since data was first collected in 1959.[30] It explained that, in the year ending March, total quarterly nominal GDP grew by over $31 billion, but just $3.1 billion of that reflected higher labour compensation, including wages, salaries and other forms of compensation such as employer superannuation contributions. In other words, less than 10 cents from every dollar in new GDP went towards boosting labour compensation. The report summed the dire situation up well: 'In short, the link between GDP expansion and workers' incomes has never been weaker.'

This trend supports the Australian Council of Trade Union's 2013 findings that there has been a significant 'decoupling' of wages and productivity since the turn of the century.[31] This situation has been worse in Australia than the US since 1995. As economist Greg Jericho asserts, 'That's a pretty damning indictment of our IR system – our workers are getting less reward for productivity benefits than the USA.'[32]

Technology has been exacerbating this trend in some tangible ways, most notably in its tendency to create 'strong monopolistic markets' for new goods and services where there is little competition.[33] For example, global powerhouses such as Google and Apple have taken advantage of market dominance and share their huge profits among a small set of employees and shareholders.

It is easier for tech giants to dominate their respective markets when they do not always remunerate the real 'owners' of their product. Facebook and many other digital businesses such as LinkedIn make money from people's personal information through targeted advertising. The millions of people who provide the platform with its content don't make a cent directly.

THREE SCENARIOS

Just as there was no clear answer to how many jobs will be lost to artificial intelligence, robotics and automation – and whether those jobs will be replaced – the same is true for the extent to which technology will worsen inequality.

In 2016 IMF researcher Andrew Berg and his colleagues developed sophisticated models to evaluate three distinct scenarios.[34] In their first approach, they assume robots to be a virtually perfect substitute for human workers akin to the human-like machines of science fiction films. In this model, output per person rises, but inequality also worsens. That is because these humanoids increase the total labour supply (people plus robots) and drive down wages. Given the robots are profitable, they attract investment away from traditional capital such as buildings and conventional machinery, which lowers demand for those working with that traditional capital. Berg and his colleagues estimate that, in this scenario, both the benefits of higher production and the harmful effects of lower wages will grow over time.

In their second model, which they describe as more likely, they assume robots are a close – but not perfect – substitute for human workers. Higher production and increased labour supply are still in play initially, but humans' unique talents – things like creativity – become more and more valuable as time goes on in this scenario because they can't be replicated by machines. Eventually, the increased labour productivity resulting from human talents being augmented and complemented by increased traditional capital and increased investment in robotics will mean wages start to rise. But that turning point is likely to take twenty years in their model and, given robots would still have a massive role in the economy, wages and equality would not be likely to return where they were in the pre-robot era.

The third scenario is the most alarming. It divides workers into two categories: skilled and unskilled. Robots are close substitutes for unskilled workers, but not for skilled workers. The model assumes that the split in numbers between skilled and unskilled workers is almost even. In this scenario, skilled workers' wages rise relative to their unskilled counterparts, and also in absolute terms, because they are more productive when combined with robots. Meanwhile, unskilled workers' wages plummet as they are pitted directly against robots.[35] Inequality is dramatically inflamed. This model estimates that, over a fifty-year period, real wages for low-skilled workers plummet by 40 per cent, and the group's share of national income drops from 35 per cent to 11 per cent.

It is concerning, even chilling, that in each scenario most of the income goes to those who own the capital and to the skilled workers who aren't at risk of being replaced by robots and automation – the only real variation is the extent to which this takes place. The authors say the fate of those left behind is clear: 'Low wages and a shrinking share of the pie.' This would be a social and economic catastrophe.

UPSIDES AND ANXIETIES

Most of this chapter has been quite grim. It is important to recognise that there are also plenty of instances where new technology has helped workers. There has been the computer, the typewriter, the tractor and, if we go back far enough, the wheel. Berg and his co-authors point to a more recent example – mapping software, which can make drivers safer and more efficient.

Technology and digital disruption can also increase the size of markets. Mohit Sharma, the director of Australian-based automation services firm Mindfields, points out how Uber grew the cab market in San Francisco from $200 million to $1 billion. Technology, he says, has always made life more comfortable and 'releases human bandwidth for more productive use'.[36]

Nobel Prize–winning economist Edmund Phelps coined the term 'mass flourishing' to describe this kind of extra bandwidth.[37] He describes two components – the growth of productivity and wages and, perhaps more important, the 'successful exercise of creativity and talents'. Phelps believes this mass flourishing first appeared during the Industrial Revolution, where dull work gave way to rewarding jobs, and people, as they used their imagination to create new things, experienced personal growth. It was not scientific advances that led to mass innovation in England, he says, but 'economic dynamism', or the desire and space to innovate.

While there are plenty of contemporary examples of the benefits of technological change, few share British journalist Paul Mason's highly optimistic views of its role in the future.[38] Mason paints a picture of a socialist utopia where automation will make work optional and inequality will come to an end. He predicts 'useful stuff' will be made with minimal human labour, shared freely and commonly owned. This includes the machines themselves. The end result, he says, is 'a world of free machines, zero-priced basic goods and minimum necessary labour time'.

We share neither Mason's optimism nor the most defeatist views. But the existing data and predictions are more than enough to concern us about the potentially negative impact on workers and inequality.

Already, that anxiety exists. *The Deloitte Millennial Survey 2017* found Australians born between 1982 and 1999 were increasingly doubtful about their futures. Forty per cent thought automation posed a threat to their jobs, 44 per cent believed there would be less demand for their skills, and more than half (51 per cent) said they would have to retrain.[39]

Barack Obama considered these concerns so prevalent in the US that they warranted a mention in his January 2017 farewell speech, which was otherwise rich with hope and positivity. 'The next wave of economic dislocation won't come from overseas,' he warned. 'It will come from the relentless pace of automation that makes a lot of good, middle-class jobs obsolete.'[40]

We cannot afford to ignore this. Tyler Cowen succinctly sums up the stark reality:

> If you and your skills are a complement to the computer, your wage and labor market prospects are likely to be cheery. If your skills do not complement the computer, you may want to address that mismatch. Ever more people are starting to fall on one side of the divide or the other.[41]

Cowen's warning is confronting, but there is a glint of hope. It comes with a choice. That choice applies to society at large as well as to its leaders. Technological change can actually be good for equality and good for individual workers, but heading down the right path cannot be left to chance; it requires dedication and foresight and our best efforts. We need to be as creative and inventive in ensuring opportunity is shared by all in the new machine age as we are when developing technology in the first place.

CHAPTER 5
IS OUR EDUCATION SYSTEM PREPARING US FOR THE FUTURE?

When Donald Horne's celebrated book *The Lucky Country* was first published in 1964, Australians were enjoying, by world standards, a very comfortable existence. While Horne's title was an accurate reflection of Australia's good fortune in having an endowment of national resources and well-functioning administrative systems inherited from Britain, it was meant to be more than a little ironic. Horne believed Australia had not earned its luck. He saw a nation in the early 1960s that was at risk of faltering in the decades ahead if it did not become more ambitious and less conservative. In his words: 'Australia is a lucky country run by second rate people who share its luck. It lives on other people's ideas.'[1]

Australia did make very important cultural, social and economic changes. The Hawke–Keating microeconomic reforms of the mid-1980s to mid-1990s helped Australia become more internationally competitive; more recently, the actions taken by the Australian Labor government during the global financial crisis meant that we did not suffer large rises in unemployment, as other Western countries did. This was neither luck nor happy accident.[2] Today, Australians are among the richest people in the world, in aggregate.

The question now is whether we can again take the initiative and make the transition to ensure that Australia remains prosperous in a technology-dominated world. This is our generational challenge. Success will not happen by accident and we cannot depend on good fortune to get us through. As in the past, we will need to make our own luck.

In their book *Two Futures*, Jim's parliamentary colleagues Clare O'Neil and Tim Watts invite us to envisage the world twenty-five years from now, when technology pervades almost every aspect of our lives.[3] They ask us to imagine two scenarios. In the first, Australia's education system has failed to prepare the workforce for the new world. Some individuals make a lot of money from their ability to use digital technology, but the majority of Australians are left behind. In the second scenario, Australian people and businesses embrace the Digital Revolution. Computational thinking is taught in schools alongside reading, writing and arithmetic. People are willing and able to take advantage of the Digital Revolution.

O'Neil and Watts conclude that the choice is ours. It depends on the decisions made by the Australian people and their governments. We can continue down the path where data and evidence are ignored; where climate change is denied; where a broadband speed of 25Mb/s is believed to be sufficient for most Australian homes for the next few decades; and where properly funding all schools is not believed to be necessary to prepare our children for the future. Or we can face up to the risks to employment and equality discussed in the preceding chapters and accept that Australians may be entering a very different world of work to that of today.

INVESTMENT IN SCHOOLS

Investing in education and training is both sensible and prudent in light of this potential risk. That is why it is both curious and disappointing to encounter such reluctance to support full and far-sighted investment in education.

Better education and training is rational not only as a national investment, but also on the grounds of equity and social cohesion, as George Megalogenis observed in his brilliant Quarterly Essay *Balancing Act*.[4] Megalogenis reported that migrants hold almost all of the full-time jobs created in Australia since 2007. He noted that the migrants didn't displace the local-born; they just took the cream of the new positions on offer, most notably in the professions:

> The post-war assumption that migrants to Australia will only do the work that the locals don't want no

> longer applies. Now migrants are also being hired for work that the locals are not qualified for.
>
> If the political system couldn't cope with the infrastructure demands of the past decade, how will it respond to a future in which a fraction of the local-born feel that they have been pushed to the margins of society? The logical answer is to dramatically increase public investment in education to ensure these local-born are not left behind.

The educational goal of Australian governments should be clear: to prepare for a future in which many routine jobs, both manual and cognitive, will be done by machines. The good news is that the introduction of sophisticated robotic and computational systems will take time – not only to get the technology working properly, but also to work through the many safety, regulatory and ethical issues that will arise. So it will likely be an evolutionary process, but one that we need to prepare for as a society.

There have been similar challenges in other times. As the Industrial Revolution progressed, Britain lost its lead to other countries, particularly the US and Germany. The generation who inherited the businesses of the original British entrepreneurs was likely to have received a 'classical' education, often with an emphasis on Latin and Greek. Such an education was not the best preparation for developing modern businesses and industries in an increasingly competitive world. By contrast, the German education system put a greater emphasis on science and engineering. Germany recognised the importance of a formal scientific education for those who would lead their emerging industries, whereas the British seemed to have a bias against it.

The lesson is clear for Australia today. We may not have our heads in the sand in the same way as the British did in the nineteenth century, but we would be wise to increase the speed at which we are changing our education system.

JOBS OF THE FUTURE

In 'The Future of Jobs', a recent report by the World Economic Forum, several of the world's largest companies were surveyed.[5] AI, advanced robotics, biotechnology and genomics were predicted to have significant impacts on jobs from about 2020. The report foresaw more immediate disruption to the workforce by cloud technologies, big data, the sharing economy and an ageing society:

> It is clear from our data that while forecasts vary by industry and region, momentous change is underway and that, ultimately, it is our actions today that will determine whether that change mainly results in massive displacement of workers or the emergence of new opportunities. Without urgent and targeted

> action today to manage the near-term transition and build a workforce with future-proof skills, governments will have to cope with ever-growing unemployment and inequality, and businesses with a shrinking consumer base.

The WEF argues that most existing, traditional education systems, with their silos of subjects, are not well placed to meet future needs. Major programs to retrain and upskill the existing workforce will be required. The report notes that technological change will create many new roles, for which workers will require a combination of technical, analytical and social skills.

By way of illustration, the WEF highlights two jobs that are important in almost all industries and geographical regions: data analysts and specialised sales representatives. More data analysts are needed to make sense of, and derive insight from, the mass of data that will be generated by new technologies. And more specialised sales representatives will be needed by every industry to explain their technology-based products to customers and users.

Similarly, the CSIRO gives six examples of jobs likely to be important in the future: data analysts, decision-support analysts, remote-controlled-vehicle operators, customer experience experts, personalised preventative-health helpers and online chaperones.[6]

As such jobs emerge in the years ahead, other roles will persist for the foreseeable future. For example, personal care and personal services roles will be important, particularly as the Australian population ages. Jobs in the creative arts or sports, where people value human creativity or talent, are likely to survive. And non-routine manual jobs, such as garden maintenance, furniture removal, car repair and most food preparation, should be relatively safe. These types of roles are all unlikely to be done substituted with robots and AI systems – at least for the next few decades. We can identify five characteristic skill groupings that are hard to replace:

- Non-routine manual skills applied in diverse environments.
- Non-routine interpersonal skills applied in diverse environments.
- Non-routine cognitive skills requiring creativity.
- Non-routine cognitive skills involving critical thinking and problem-solving where there is no predefined rules-based procedure.
- Non-routine cognitive skills requiring the acquisition and analysis of new information.

It is fairly straightforward to map these skills onto many new and existing jobs. For example, the specialised sales representative mentioned above would likely need excellent interpersonal skills, the ability to analyse new information and a level of creativity to translate the product or service to fulfil customer needs or desires. Gardening, on the other hand, largely requires the exercising of non-routine manual skills and occasionally solving unstructured problems.

As we have seen, it is not the case that all future jobs require a mastery of technology or technical skills. The UK Forum on Computing Education has defined four categories of digital competence:

- Digital Muggle: No digital skills – digital technology may as well be magic.
- Digital Citizen: Able to use digital technology purposefully and confidently to communicate, find information and purchase goods and services.
- Digital Worker: Includes the ability to evaluate, configure and use complex digital systems. Elementary programming skills such as scripting are often required for these tasks.
- Digital Maker: Has the skills to actually build digital technology, typically involving software development.[7]

The strategy and advisory company AlphaBeta used this categorisation in a 2015 report analysing the digital skill requirements of 405 Australian occupations.[8] The following diagram, reproduced from the report, projects that more than half of Australian workers will, in the period 2017 to 2020, need to be operating at the Digital Worker or Digital Maker level.

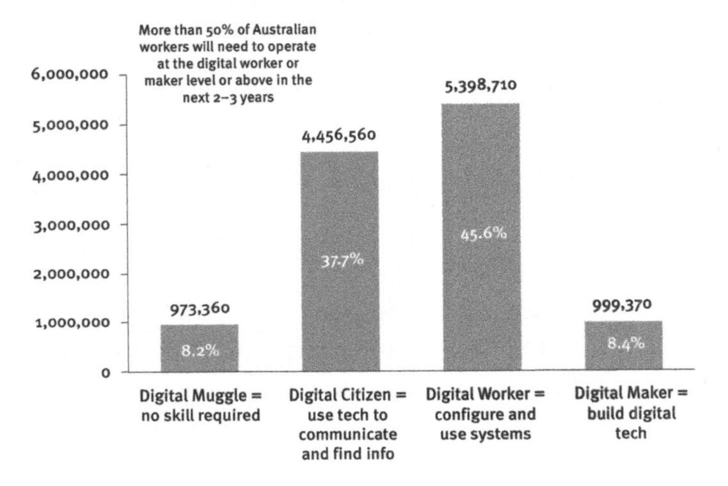

Source: ABS, UK Digital Taskforce, AlphaBeta analysis.

When a student is in primary or high school, no one knows with any certainty what jobs – even what types of job – they will have throughout their working life. So for reasons of both practicality and fairness, it is imperative that all students are given the opportunity to learn the basic skills that prepare them for the technologies they may meet later in life. As the AlphaBeta report notes:

> The Australian workforce also needs around 5.4 million 'digital workers'. These will not necessarily be STEM employees, although some will be. Instead, they are workers who understand how to make technology work for them. To grow this skillset,

> it is worth considering a mandatory computing curriculum in primary and secondary schools.

People already in the workforce are quite likely to experience a change in employment, perhaps a quite radical shift – from manufacturing or construction into aged care or healthcare, for example. This type of transition may take them from a role in which a familiarity with technology is not important to one in which it is an integral part of the job.

There is also a growing need to retrain and upskill Australia's existing workforce, given that more than half of the jobs in the near future are going to fall into the categories of Digital Worker or Digital Maker. (This issue will be taken up in the next chapter.)

We cannot ignore that computation is becoming embedded in many disciplines, from agriculture to biochemistry to cartography to marketing to zoology, and is indispensable for data gathering, analysis and optimisation. Many Australians realise that if we do not keep up with, and master, these computational technologies and the disciplines that underpin them, we will be at a competitive disadvantage, particularly compared with workers in rapidly developing Asian countries.

This has led to much discussion and concern about the teaching of the STEM subjects: science, technology, engineering and mathematics. Mathematics is rightly seen to be a foundation for most sciences and engineering, which generates much of the wealth and prosperity of modern economies. Mathematics is also fundamental to the emerging technologies of AI and robotics, as well as many other technologies, so it is not surprising that our performance in international testing of mathematical competence receives a lot of media attention. While today, on international comparisons, Australian students may not be performing badly, the trends are heading in the wrong direction.

We have an opportunity to reverse our declining performance in this area. But we need also to be realistic and acknowledge that only a minority of people will need a knowledge of advanced mathematics for their job. For Digital Workers or Digital Makers, though, it is questionable whether the traditional way of teaching calculus, for example, is well suited to their future needs. They will almost certainly have powerful tools capable of solving more complex problems than traditional calculus classes could possibly cover. But to use these powerful tools, the students of today will need to understand the fundamental underlying concepts. The focus for mathematics students today should be on how to formulate a problem in a way that a machine can perform the often complex but tedious calculations, and then how to understand and interpret the results of those calculations. Repeated exercises in actually performing those calculations is of little value to today's students.

Even a Digital Muggle, whose job requires very little or no digital technology use, will still require a certain level of competence in their everyday life – for social or educational needs, for example, or communication. It is becoming increasingly difficult to navigate today's world without some level of competence with technology. And certainly a better appreciation of statistics would greatly assist many citizens in evaluating fact from fiction when claims are made regarding scientific issues, including climate change or medical issues such as vaccination.

For both future work requirements and informed participation in social and economic debates, it is important to understand mathematical/statistical concepts and the methods used by modern scientists in evaluating data. There needs to be a large-scale effort to provide students with 'computational thinking skills' – the thought processes required to translate a problem into a form that a computer can deal with and then interpret and communicate the output of the computation to provide an effective solution to the problem.[9] We'll discuss computational thinking in more detail later in this chapter.

STEM SUBJECTS

The trends in mathematics and science enrolments over the last few decades have received a lot of attention in the media and among policymakers.[10] In 2016 an analysis done by the Australian Mathematical Sciences Institute (AMSI) found that for Australian Year 12 students:

- Almost 20 per cent were not enrolled in any mathematics course in 2015.
- More than 50 per cent were enrolled in the elementary (non-calculus) level of mathematics in 2015, an increase from 37 per cent in 1996.
- The percentage of students enrolled in intermediate mathematics dropped from 29.6 per cent in 1996 to 19.2 per cent in 2015.
- The percentage of students enrolled in advanced mathematics dropped from 13.6 per cent in 1996 to 9.6 per cent in 2015.
- There is a significant gender bias for advanced mathematics, with the ratio of boys taking advanced mathematics to girls being almost two to one.[11]

The picture for Australian high school science is, according to the AMSI report, not quite as grim as for mathematics but nevertheless shows disturbing long-term trends. Participation rates, based on Year 12 enrolments for physics, declined from 21 per cent in 1992 to 14 per cent in 2012. For chemistry, the decline was from 23 per cent to 18 per cent, and for biology from 35 per cent to 25 per cent. For a country with a focus on innovation, this is not an encouraging picture.

One factor that may have contributed to these disappointing trends, especially in mathematics, is changes to university entrance requirements. Many universities eliminated the requirement of intermediate or advanced mathematics for admission to science and engineering degrees. AMSI reports that only 14 per cent of universities set a prerequisite of a minimum of intermediate mathematics for entry into a science degree. The requirements for engineering degrees are a bit more stringent, with 59 per cent requiring a minimum of intermediate mathematics for entry. Fortunately, some universities are reversing their position and reinstating mathematics prerequisites for a number of their degrees.

Another possible reason for these trends may simply be that a greater proportion of students are now continuing to Year 12. These extra students are likely to be less academically inclined and hence more likely to enrol in either a lower-level maths course or none at all.

A 2014 paper by three Australian academics comes to similar conclusions as the AMSI report.[12] The paper places Australian Year 12 high school mathematics courses into three categories – Entry, Intermediate and Advanced – with the Entry-level courses being non-calculus based. The paper finds that 'students of both sexes are abandoning Advanced mathematics in favour of Intermediate, while the changes from Intermediate mathematics sees male students selecting Entry mathematics while females elect no mathematics'.

It also appears that, at least in New South Wales, there may be another factor at work. David Pitt from Macquarie University notes that:

> Recently, a number of high school teachers, students, and academics across NSW have questioned the suitability of the scaling process used in the calculation of the ATAR [Australia Tertiary Admission Rank], which is the measure used in the admission process to Australian universities ... Of more than 1000 mathematics teachers surveyed, 51% believed that some students in their school were selecting senior mathematics courses below their capability. A desire to optimise HSC and ATAR results was the most common reason given.[13]

The suspicions of these NSW maths teachers were supported by Pitt's statistical analysis. NSW students correctly believe that the algorithm used to scale the maths results for the ATAR rankings gives an advantage to those enrolling in the lowest level. Not surprisingly, they make what seems a sensible subject choice regarding the level of mathematics that will maximise their marks.[14]

TEACHERS AND TEACHING

How do we address these troubling trends and reverse the declining popularity of mathematics and science in senior high school? It is now accepted that technology is enabling globalisation, which means that we are in a global competition for highly skilled and high-paying jobs. Andrew Leigh, for example, argues:

> Education remains the best productivity-boosting policy – and the best antipoverty vaccine – that we have yet developed. To transform teacher effectiveness would both raise the rate of innovation and entrepreneurship, and reduce joblessness and inequality. If the robots are coming, we'd best ensure we have great teachers to meet them.[15]

Helping existing science and maths teachers become more effective is one of the aims of the In2Science program, a collaboration between four Victorian universities – La Trobe University, RMIT, Swinburne University of Technology and the University of Melbourne. These universities place volunteer university-student mentors in secondary school classrooms for ten weeks. As well as assisting teachers, they provide a role model for students, encouraging more of them to study maths and science through to Year 12 and beyond. They also provide real-world examples, for both students and teachers, of how mathematics and science are used to solve problems.

Since the program started in 2004, close to 2000 mentors have been placed in 140 Victorian schools, reaching more than 50,000 students. A similar model will be introduced in Queensland primary schools from 2018.[16] Such programs also transfer knowledge to teachers, building confidence and resulting in an increased number of inspirational teachers – those teachers who have a positive influence during a student's formative years. For all these reasons, training and mentoring more STEM teachers is our first recommendation in this chapter.

The teaching model that has been in existence for many decades is undergoing a progressive transformation. The new model is referred to as 'flipped learning', and expanding trials of this format is our second recommendation. A pilot program involving nine UK schools showed that most teachers saw the flipped learning approach as valuable – but as one approach among a wide repertoire of teaching strategies.[17] The material the student needs to learn can be made available as an application or in video format and is usually watched outside the classroom. Students come to class with an exposure to the new material and a preliminary understanding of the topic. If designed well, these types of applications engage the student while also providing teachers with feedback on the exercises done, how well the student is performing and where they are having difficulties. The teacher can then spend more time answering questions, clarifying concepts and coaching students individually or in small groups. It also puts more emphasis on students taking responsibility for their own learning.

The approach requires teachers to plan how they will use the lesson time that is no longer needed to cover the standard material. One possibility would be to focus on imparting some of the skills and capabilities that are going to be increasingly important for students to master – such as critical thinking, cooperative problem-solving, creativity and the ability to evaluate the reliability of data sources.

This, of course, means that teachers are going to need to be trained in how best to teach these skills. For teachers who graduated many years ago, this is likely to be a major change from their original training. Hence the need to invest in ongoing teacher training and in the hardware and software needed by schools to support these new types of teaching models.

Not only are teaching strategies and methods changing, so too are the content and skills to be taught. For example, the past few years have seen considerable efforts made in some Australian states to include probability and statistics in some of the intermediate and advanced mathematics courses. Given the important role that statistics plays in the natural, social, biological and medical sciences, this is a good thing.

Likewise, it is encouraging to see the Queensland Curriculum and Assessment Authority developing a number of new Year 11 and 12 subjects. The draft syllabus for one of the new subjects, 'Digital Solutions', specifies that students will learn about algorithms, code and user interfaces through generating digital solutions to simple problems. The aim is that they will also learn creative problem-solving, critical thinking, effective communication and collaborative skills. Unfortunately, this is not a course that will be undertaken by all or even a majority of Queensland high school students. Our third recommendation is that this curriculum, or one like it, should be adopted nationwide.

The Queensland government also plans to make coding and robotics compulsory in all state schools – primary and secondary – by 2020.[18] Extending this to the entire nation is our fourth recommendation.[19]

Our fifth recommendation is the inclusion of computational thinking in the teaching of maths and science in all Australian schools. This would guarantee that all Australian students would be exposed to this important subject, and may well motivate more Australian students to enrol in maths and science at higher levels.

COMPUTATIONAL THINKING

The term 'computational thinking' was first used in 1980 by the South African–American mathematician, computer scientist and educator Seymour Papert. The term gained greater prominence in 2006 when Jeannette Wing proposed in the *Journal of the Association of Computing Machinery* that computational thinking become an essential part of every child's education. In 2012 Wing defined computational thinking as the thought processes involved in formulating a problem, and expressing its solution, so that it can be effectively dealt with by a computational entity.[20] Subsequently, educationalists have worked to define the concept more precisely and provide practical steps for its teaching.

Stephen Wolfram, the British-American physicist, mathematician and computer scientist, says that computational thinking 'is about formulating things with enough clarity, and in a systematic enough way, that one can tell a computer how to do them'.[21] But to tell a computer how to do something, you need a programming language. Not surprisingly, but with some justification, he believes that his own Wolfram Language, a knowledge-based programming language, is an ideal vehicle. He writes:

> It's programming in which one is as directly as possible expressing computational thinking – rather than just telling the computer step-by-step what low-level operations it should do. It's programming where humans – including kids – provide the ideas, then it's up to the computer and the Wolfram Language to handle the details of how they get executed.

The International Society for Technology Education and the US Computer Science Teachers Association provide a list of 'dispositions and attitudes that are essential dimensions of computational thinking'.[22] These include:

- Confidence in dealing with complexity.
- Persistence in working with difficult problems.
- Tolerance for ambiguity.
- The ability to deal with open-ended problems.

- The ability to communicate and work with others to achieve a common goal or solution.

There is scope to apply computational thinking across many areas of a school curriculum, at both primary and secondary levels. Students could, using computational tools that are now readily available, create a social network graph of the characters in a drama, for instance, or download images of paintings and analyse the colour distributions used by different artists over time, or analyse how verb structure is related to the origins of different languages. Undertaking a social survey aided by computational tools would require a student to employ computational thinking as they work through the survey design and analyse results. These tasks are not about the mechanics of programming or coding, but instead the bigger question of how to think about and structure the overall problem. This allows the problem or question to be put into computational form so that the power of modern computational tools can be applied.

It is important to draw a clear distinction between coding and computational thinking while also understanding the relationship between them. 'It's a little bit like the relation of handwriting or typing to essay writing,' Wolfram explains. 'You (normally) need handwriting or typing to be able to actually produce an essay, but it's not the intellectual core of the activity.'

Given the proliferation of programming languages, some of which occupy specific niches in terms of the types of application for which they are best suited, it is important to ensure that when teaching students computing skills the coding language chosen is quite general. The aim should be to promote an understanding of computational principles rather than the specifics of a particular programming language.

There are some parallels between computational thinking and mathematics, as both require clarity of thinking and a level of precision in expressing concepts and relationships. These parallels make mathematics courses an obvious place to begin teaching computational thinking. There is real potential to make mathematical concepts 'come alive' when students explore ideas using a powerful computational tool and solve problems using both mathematical and computational thinking. A deeper level of understanding of mathematical concepts may be the result. As the famous American mathematician and computer scientist Donald E. Knuth puts it:

> It has often been said that a person does not really understand something until he teaches it to someone else. Actually a person does not really understand something until he can teach it to a computer, i.e., express it as an algorithm … The attempt to formalize things as algorithms leads to a much deeper

> understanding than if we simply try to
> comprehend things in the traditional way.[23]

A recent paper by David Weintrop and his colleagues breaks down computational thinking into four main categories: data practices, modelling and simulation practices, computational problem-solving practices, and systems thinking practices. Each of these categories is further decomposed. For example, in the data practices category, there are skills such as collecting, manipulating, analysing and visualising data. In the systems thinking practices category, there are skills such as viewing a complex system as a whole, understanding the relationships between entities within a complex system and being able to effectively communicate information about such systems.[24]

This type of knowledge with its associated skills is not yet part of any formal curriculum widely implemented in Australian schools. Weintrop and his colleagues argue there are several good reasons for making computational thinking an integral part of maths and science courses. First, as outlined above, there is a complementary relationship between computational thinking and the teaching of maths and science. Students will quickly learn that using such tools forces a deeper understanding of the underlying mathematical or scientific principles. Most computational tools are unforgiving if they are used without having fully grasped the problem or concepts that underlie a solution, whereas many of the existing simple aids, such as calculators, do not force this deeper level of thinking about the problem and its solution.

The advantages also flow the other way: science and maths provide excellent problems for applying computational thinking. Having authentic problems with real-world application can be highly motivating for students and help counter the bane of most mathematics teachers' existence: the question 'What is the use of this stuff?'

Second, if computational thinking were included in mathematics and science courses, there are already many proficient teachers who, with appropriate training, would be well placed to include these concepts and practices into their lessons.

Third, the fields of science and engineering have undergone major transitions due to new analytical approaches and the availability of sophisticated computational models and ever more powerful computers. So including computational thinking in science and maths lessons brings them closer to the way professional scientists and mathematicians now work. Also, many of the computational thinking tools used in mathematics and science courses are in fact the same tools that professional scientists and mathematicians use. This potential convergence between the educational setting and the professional workplace can give students a head start and is also likely to be highly motivating.

The knowledge and skills our education system provides to our children will largely determine the opportunities that are available to them. This, in turn, will have a bearing on whether they realise their potential, and on their future wellbeing. While this has always been true to some extent, it is becoming even more of an issue in an increasingly globalised and technology-dominated future. And while machines will progressively encroach on existing jobs, for the foreseeable future there will be many jobs, old and new, that computers and robots will not be capable of doing – or at least not without input or oversight by humans. Some of these jobs will not require significant formal education, but a significant number will increasingly require a familiarity with, and in many cases a mastery of, the skills and practices that are part of computational thinking.

In their 2013 report, Frank Levy and Richard Murnane of the think tank The Third Way concur: 'For the foreseeable future, the challenge of "cybernation" is not mass unemployment but the need to educate many more young people for the jobs computers cannot do.'[25]

The job of ensuring that young people receive an education that equips them for the future falls largely on governments. Their role, particularly in public education, is to provide 'equality of opportunity' to all children, irrespective of parental incomes and home environment. That is why our sixth recommendation, proper and full needs-based funding for schools, is so important to combat technological inequality.

Levy and Murnane also highlight the importance of the school system in providing mathematical skills:

'While children learn much of their vocabulary from the home and environment, children learn most of their mathematics in schools, giving schools more leverage in developing students' mathematics skills.' The same is likely to be true for computational thinking skills.

EDUCATION, SOCIAL MOBILITY AND EARLY YEARS

Mathematics, science and computational thinking are, of course, not the only conceptual tools needed in the workplace of the future, or to understand the social, economic and political issues that confront Australian citizens. Much of what we know, or think we know, comes from media of one sort or another. So the ability to distinguish high-quality and largely unbiased sources of information from poor-quality sources that may be pushing an agenda is obviously important. As is the ability to analyse, deliberate upon, summarise and draw valid conclusions from information and data.

These same capabilities are equally important in many work settings, together with some of the 'softer' skills and abilities such as emotional intelligence, communication and presentation skills, problem-solving and creativity of various kinds. It is sobering to think that teachers at every level, from primary school through to university, will need to impart these types of skills.

Focusing on the early years is critical. This focus, especially in low socio-economic status (SES) schools, is our seventh, and final, recommendation in this chapter.

Economist Miles Corak's 2013 paper on inequality and immobility examines the effect of investment in education on inequality.[26] He argues that if it is targeted at early childhood, then the investment in education will reduce inequality, but if it is largely targeted at later years, especially university education, then it is more likely to increase inequality. Corak points to the US as an example, since it spends more on the schooling of its children than any other high-income country. But for every dollar spent on primary education, three are spent on tertiary education. If the disadvantaged fall behind in the early years, they are more likely to drop out of the educational system.

This is why it is so important to ensure that all children, irrespective of their family circumstances, have the opportunity to master the basic skills of reading, writing and maths. To this list we would now add computational thinking.

TECHNOLOGY AND PARTICIPATION IN SOCIETY

No one can be certain what the next decades will bring – technology is evolving quickly. But whichever of the scenarios we have discussed plays out, there is no downside to doing everything possible to provide Australians with the knowledge and skills they need to interact with technology at whatever level is appropriate for them – whether they design such technologies, apply sophisticated technology in their own domains of expertise or use technology casually in or outside their place of work.

To summarise, the specific recommendations we have proposed in this chapter are:

1. Training and mentoring more STEM teachers.
2. More, and more substantial, 'flipped learning' trials.
3. Adoption of Queensland's 'digital solutions' curriculum.
4. Compulsory coding and robotics in primary and secondary schools.
5. Emphasis on computational thinking in existing subjects.
6. Needs-based funding for schools to combat technological inequality.
7. Early education and intervention, especially in poorer communities.

Fostering engagement and encouraging people to participate in our democratic processes is now an important priority for most democratic governments. Individuals now need to evaluate the decisions their governments are making on sometimes highly technical areas such as climate change, biotechnology, AI and vaccination.

So an understanding of the concepts underpinning the mathematical and computational sciences will allow people to become more informed and active citizens. It will also allow them to benefit from new technologies. It is a sad reality that those people in our society who today cannot use the internet are at a distinct disadvantage in almost all spheres of life. This capacity broadens people's lives and provides them with a better appreciation of the world in which they live – not just the natural world, but also the world resulting from the technologies that are now being created.

It may be tempting to say that this aspiration is not realistic, that not everyone is going to be interested in maths, science and computation. But we would not accept this argument with respect to reading and writing. We know it is impossible to participate effectively in today's society without literacy skills. We perhaps need to accept that it will be impossible to participate effectively in tomorrow's society without some skills in maths, science and computation. We use writing skills to communicate with other people, but in a world where technology and machines play an increasingly important role, we also need to be able to communicate effectively with machines.

We owe it to our children to provide them with the knowledge and skills that give them the best chance of a successful, engaging and purposeful life. No doubt, achieving this will not be easy. But perhaps we can take inspiration from Professor Michelle Simmons' 2017 Australia Day address:

> I am grateful for that Australian spirit to give things a go, and our enduring sense of possibility. In this, we have so much to be thankful for – and, more importantly, so much to look forward to. But there is room for improvement as well. In our innovation policies, in our education system, and in the ambitions of our scientists and discoverers, I want Australians above all to be known as people who do the hard things.[27]

In the Hawke–Keating era, it was only through government, businesses and unions working together that very difficult reforms were planned and executed. The same will be true of Australia's adaptation to a technology-dominated future, particularly when it comes to our education system. It would be wise to avoid the errors made by the British education system during the Industrial Revolution. We need to be looking forward, not just in our schools but across every aspect of government policy.

CHAPTER 6
WHAT GOVERNMENTS COULD DO

Six-year-old Elizabeth Bentley arrives at the mill at 5 a.m., dreading another long and exhausting day of work. Her job requires her to replenish the rows and rows of spinning frames, which run all day. Despite being allowed only a forty-minute break before she heads home at 9 p.m., she knows she can't afford to succumb to fatigue. If she can't keep up with the machinery, her supervisor will be quick to deliver a blow with his strap, the sting of which she has felt countless times before.

This is what 23-year-old Elizabeth told a UK parliamentary inquiry in 1832 about her experiences as a child labourer, which eventually left her crippled and unable to work.[1]

It is even more depressing to realise that her experience was the norm – and far from the worst – in early-nineteenth-century England at the height of the Industrial Revolution. But it was investigations such as the Sadler inquiry, in which she gave evidence, the writings of intellectuals and the activism of progressive groups which led to legislative change in the form of the Factory Acts.

These acts put an end to the oppressive and abhorrent conditions faced by workers. They limited the hours that could be worked by women and children; led to the introduction of a ten-hour work day, which was criticised by opponents at the time; and formed the basis for modern protections that are often taken for granted today, such as a decent minimum wage and fair working conditions.

It was only through legislative intervention that the rights of workers and the most vulnerable in Industrial-era England were protected. The changes helped to ensure that it was not just the factory owners – the owners of the machines – who were reaping the benefits of the technological change, but the wider community too.

Former US Treasury secretary Larry Summers cites leaders such as William Gladstone in the UK and former US President Theodore Roosevelt as examples of those who embraced technology-led change instead of fighting it.[2] In doing so, they helped ensure technology's harshest impacts were softened and its benefits were distributed fairly. For Gladstone, it was universal primary education, and for Roosevelt, a progressive agenda from antitrust laws to utility pricing. Their hands were strengthened by the extension of the franchise to give voice to workers in the factories as well as landowners.

The lessons for today's technological revolution are clear. We need leadership, foresight and democratic licence, and we need to care about how the impact of change is distributed. Getting it right means harnessing the enormous upside of change to benefit the community at large. Getting it wrong means more people on the employment scrapheap, and more and more benefits of the machine age flowing into fewer hands at the top.

THREE PATHS

Chapter 4 argued that worrying trends toward greater inequality have put governments on notice. They have policy choices to make that do not always accord with the traditional left–right political divide. And, like Goldilocks, they have three options.

The first approach, to allow change at any cost by doing nothing, is irresponsible and dangerous. It relies on a notion we rejected earlier – technological trickle-down – allowing the owners of the machines to reap the massive gains of technological change in the hope that some of the benefits might then flow down to the rest of society. Like the long-discredited idea of trickle-down economics, technological trickle-down only serves to exacerbate and accelerate inequality.

As prominent academic and Microsoft computer scientist Kentaro Toyama's writings make clear, 'technology is only an amplifier of human conditions'.[3] In other words, in a highly functioning society or organisation, technology will make humans more efficient. But in a poorly functioning society, technology will amplify the existing negative dynamics. In this sense, simply to 'let it rip' would only guarantee that the impact of technological change is unfairly distributed.[4]

The second approach, to reject change or try to hold back the tide, is just as unhelpful. For some time, resisting technological progress has been synonymous with the plight of the Luddites, textile workers who, in protest, destroyed weaving machines during the Industrial Revolution in England. Today's Luddites wrongly believe that technological change can't be a force for good, that we don't have the capacity to harness it, and that we can prevent it.

Consider American writer Nicholas Carr, one of automation's most outspoken critics.[5] Not only does he argue machines jeopardise safety by leading to skills deterioration – he gives as an example pilots who rely too much on autopilot – he also doubts governments can address inequality caused by severe changes to the labour market:

> As history reminds us, high-flown rhetoric about using technology to liberate workers often masks a contempt for labor. It strains credulity to imagine today's technology moguls, with their libertarian leanings and impatience with government, agreeing to the kind of vast wealth-redistribution scheme that would be necessary to fund the self-actualizing

> leisure-time pursuits of the jobless multitudes. Even if society were to come up with some magic spell, or magic algorithm, for equitably parceling out the spoils of automation, there's good reason to doubt whether anything resembling the 'economic bliss' imagined by Keynes would ensue.[6]

Carr's is a minority view but it can't be ignored. Nor can the simplistic populist rhetoric making waves in many developed nations including ours, sparked, in part, by the effect automation has already had on manufacturing jobs. This approach fails to recognise that not only is it futile to try to halt technology's march, but also that extreme government intervention or restrictions could stifle creativity and innovation. That would have problematic consequences for our society and economy.

The third approach is to embrace technological change but do what we can to ensure it works for people and not against them. It means rejecting what academics call 'technological determinism' and recognising that, while these trends are big, we have agency to shape their impacts, and that our decisions about how to respond to technological change should be informed by our values.[7] Above all, it means acknowledging the extraordinary capacity of technology to improve lives while cherishing our country's egalitarian heritage and seeking to make inclusive economic growth a key feature of our future, not just of our past.

Ours is a no-regrets approach to securing a 'fair go' in the new machine age. It leans heavily on policies that are three-dimensional: forward-looking, upward-climbing and outward-facing. These are: anticipating the trends and needs of the future, not just the present; caring about aspiration, ambition and social mobility, and not stifling them; and resisting the protectionist and populist urges to turn inward.

CHOOSING THE RIGHT PATH

Governments around the world are turning their minds to the threat of automation. When he was US president, Barack Obama commissioned a report which was released in December 2016.[8] The *Artificial Intelligence, Automation, and the Economy* report contained three broad strategies: invest in and develop artificial intelligence to harness its benefits; educate and train citizens for the jobs of the future; and aid workers in their transition between jobs and empower them to share growth more broadly.

This is a more modern version of some longstanding policies, such as Denmark's 'flexicurity' model, which helps retrain workers in the face of mass job losses.[9] The Danish government insists that a 'golden triangle' of policies work together: generous unemployment benefits; massive investments in jobs, education, training or career guidance for the unemployed; and more flexible rules for hiring and firing during the peaks and troughs of the business cycle.[10]

Other countries are trying different methods. Finland is trialling an unconditional universal basic income (more on this later), and German and Swiss apprentices split their time between study and practical work experience in large firms, with a focus on future skills.[11]

Australia lacks an equivalent toolbox of approaches, and that cannot continue. We are kidding ourselves as a nation if we think our education pathways, social safety net, tax and industrial relations frameworks can stay the same in the face of rapid technological change without adverse consequences.

Already the lack of a future roadmap is being noticed, and policy makers are being called upon to take the initiative. As the McKinsey Global Institute puts it:

> Policy-makers should embrace the benefits from productivity growth and put in place policies to encourage investment and market incentives to encourage innovation. Also rethink education and training, income support and safety nets, as well as transition support for those dislocated.[12]

Chapter 5 concerned itself with educating our young people. This chapter is focused on working-age Australians: what to do with those displaced and disconnected from work; how to engage with lifelong learning and especially reskilling; how to maintain the best, most cost-effective social safety net that insures against the ups and downs of regular workplace disruption; how to update and upgrade the industrial relations regime to recognise uneven and unconventional work patterns; how to best collect taxes in a very different economy; and how to measure and monetise the real contribution real people make to the profits of big technology companies.

Each of our recommendations is geared towards meeting one or more of what we consider to be three key, broad, longstanding objectives: economic growth that is inclusive, work that is properly rewarded, and a safety net for those left behind.

The twenty broad policy directions we suggest are not for this term of parliament or even for the next; they are more like twenty ideas for the next twenty years. They are additional to the seven education recommendations in the previous chapter, and will be supplemented by another six recommendations for individual action in the next chapter.

SOCIAL SECURITY

Let's begin with what not to do: a universal basic income, paid to everyone – enough to get by on, but not to live comfortably. UBI has become *the* go-to policy for those seeking a policy response to the hollowing-out of the labour market in the new machine age, including tech moguls, academics and even some governments. The fact that this radical idea is even being considered shows just how serious and widespread concern is about the impacts of a drastically shifting labour market.

For someone with the imagination and resolve to make electric cars mainstream, send rockets into space and plan for the colonisation of Mars, billionaire and tech entrepreneur Elon Musk is surprisingly simplistic and defeatist when it comes to the impact of technology on the labour market. He has said: 'There is a pretty good chance we end up with a universal basic income, or something like that, due to automation. Yeah, I am not sure what else one would do. I think that is what would happen.'[13]

He is not alone in assuming there won't be work for a large proportion of the community. Prominent American futurist Martin Ford, in his award-winning book *The Rise of the Robots: Technology and the Threat of a Jobless Future*, argues for a UBI in the United States.[14] In his view, it is more straightforward and efficient to implement, with lower administrative costs, than targeted welfare. Ford also argues that a UBI encourages people to find work (in a 'jobless future'?) and avoid a 'poverty trap' that provides little incentive to move from means-tested welfare to a minimum-wage job.

Unlike the US and other comparable countries, Australia already has the best-targeted means-tested social safety net in the world.[15] If Australia were to adopt a form of UBI, it would be a backward step. It would mean giving the same amount of government support to a high-level CEO as to a single mum struggling to keep food on the table. It would mean dismantling a system that ensures support goes to those who need it most. If it replaced the current system, it would actually *increase* inequality, not decrease it, by substituting an untargeted system for a targeted one.[16]

A UBI would go against the notion of the fair go, which is ingrained in Australia's national psyche and crucial to our success. Especially if it was unaccompanied by changes in the broader tax base, given that the tax and transfer systems are considered two parts of one entity.

The costs would also be prohibitive. Respected social security economist Professor Peter Whiteford estimates that a UBI set at the level of the age pension and paid to every Australian adult would cost around $360 billion a year. To put that in perspective, the entire 2017–18 social services and welfare bill is less than half that, at around $164 billion. If income taxes were relied upon to fund the payment, the top personal tax rate would be pushed to 70 or 80 per cent.[17]

An even greater problem with the UBI concept is that it is too static. It doesn't adequately focus on the transitions and bottlenecks created by rapid technological change in workplaces. It would reduce the incentives for people to work and invest in themselves. In its most extreme and simplistic form, it could assume that we go straight from a job for life to no jobs at all.

It's true that some who lose their jobs to technological change will not re-enter the workforce, and our responsibility is to guarantee them a decent income. Having a social safety net to establish a minimum living standard, below which no one in our community should fall, will be more important than ever when technological change reshapes the world of work and wages. Already there are substantial concerns about the adequacy of unemployment and other benefits, and these will get worse over time.

But perhaps the biggest problem is that welfare without work doesn't guarantee happiness. Work is about more than earning money. It's about making a meaningful contribution, about teamwork, motivation for education and self-improvement. Paying people not to work can have adverse social consequences. A world of a few billionaires owning technology businesses and billions of others receiving 'sit-down money' would be a dystopian one.

We propose three improvements to Australian social security. First, big data and matching technology should be used not to demonise recipients or treat people as guilty of welfare fraud until proven innocent, but to ensure the social security system is responsive and working for people, not against them. This could involve some sort of top-up technology to take into account irregular work patterns, hours and wages, responsive in real time and better at combining some social security and some work. It would harness technology not solely to improve the budget's bottom line but to improve lives, with timely and automatic positive interventions when needed. In essence, it could provide a different safety net – a 'safety web' – by predicting social problems at the household level before they emerge.

Related to this, we need to focus on the opportunities that technology can provide for people with disabilities, enabling them to function and communicate more effectively. This will mean they will be able to work at least some of the time, and we need social security payments to factor in the changing nature of work, who is working and what their caring responsibilities might be. Our second recommendation is to think more about the possibilities of the caring services, and especially ways technology can augment the human contribution without replacing it, with digital assistants and the like.

Third, we need to reconsider the important concept of reciprocal obligation for the new machine age. We need to concern ourselves with the happiness and wellbeing of those permanently displaced as much as with their material income. We need to recognise that unemployment can damage physical and mental health, particularly among vulnerable people, and that it is concentrated among older, male workers and marginalised groups.[18]

If the future of the labour market does require fewer human hours or substantial pockets of unemployment, we'll need measures to ensure life is meaningful and rewarding for those without work. Australian futurist Ross Dawson sums up the problem well:

> This is perhaps the biggest challenge. Beyond avoiding an excessive polarization of income, we need to provide opportunities for people to tap and express their human potential, and be recognized for this. A divide in feelings of social worth could be as devastating as one of wealth.[19]

Reciprocal obligation for the new machine age will need to factor in how to help people who can't return to the workforce to live a meaningful life, either through volunteering or other pursuits. Any such model would have a different objective to the work-for-the-dole scheme, which has become a narrow and limited symbol of reciprocal obligation. As RMIT Professor Judith Bessant has argued, work for the dole was built on the misguided notion that unemployed people, particularly the young, were to blame for their own unemployment due to their own 'passivity, laziness, lack of discipline and skills'.[20] This misdiagnosis ignores the way technological change, particularly labour-linking and labour-saving technologies, can displace large swathes of the workforce. That's not to mention a more recent analysis showing the latest iteration of the scheme has done little to help jobseekers find work.[21]

Reconsidering reciprocal obligation with an emphasis on volunteering, caring and mentoring is an opportunity to turn a punitive program into a positive and rewarding one for those displaced by machines.

INCOME SMOOTHING

The shift away from life-long employment has been occurring for some time. Workers are now only staying in jobs, on average, for about 3.3 years. Social research group McCrindle worked out that, on current trends, a school-leaver in 2014 would have seventeen different employers and five completely different careers in their lifetime.[22] This phenomenon is likely to speed up as jobs are replaced, displaced and augmented by technology.

In this environment, options for 'income smoothing' need to be considered. Income smoothing means evening out people's income over time so they are less susceptible to the peaks and troughs of employment. Some proposals suggest supplementing income once a displaced worker is forced to accept a far lower-paying job; others 'top up' support during this transition to a new role. There is also the private-sector model of insurance companies' income-protection packages, though premiums are high because people are more likely to take them out when their employment is already precarious.

A more radical proposal (sometimes associated with the Australian Council of Trade Unions) envisages people or even larger employers paying a premium to the government in anticipation of episodic periods of unemployment in the new machine age. The government would then pay displaced workers a certain percentage of their previous salary for a set time. Alternatively, perhaps a future government could consider other incentives for workers to take out wage insurance, just as people today are encouraged to acquire health insurance when they earn over a certain amount.

Each of the existing income-smoothing proposals has limitations, and we are not prepared to recommend one over another. But we do propose a future government consults extensively on income smoothing and considers affordable and responsible measures for temporarily displaced workers so they don't suffer a substantial drop in living standards.

While all four policy directions above focus on social security and reciprocal obligation for the jobless, the fourth (income smoothing) specifically addresses the transitions between jobs. In addition to guaranteeing entitlements for those laid off, employment services need to keep up with much more regular displacement of workers and seek to return them to work as quickly as possible. Such programs have a chequered history in Australia, but more sophisticated skills matching techniques are now available.

SKILLS MATCHING

It's important to distinguish between different types of unemployment – cyclical, frictional and structural. The first is a product of economic downturns, while the second always exists in the economy as people move between jobs and, although it can lead to extended unemployment, is usually short term. It's structural unemployment, which is brought about by fundamental shifts in the economy, that concerns us the most. This occurs when entire industries decline, seeming to leave specialised workers without demand for their skills.[23] It essentially becomes a problem of matching people with jobs, and matching problems can be addressed by better policy. We should seek a better understanding of the skills possessed by workers in declining industries, and find better ways to match these skills to jobs, including by providing 'micro-training' support.

So skills matching for displaced workers is the fifth policy direction governments will need to consider as technology transforms workplaces. Given that so many jobs in their current form may disappear, there is merit in looking at workers' roles at a granular level and considering what skills might be transferrable. For instance, there may not be value in helping a truck driver find another job as a truck driver if that role has become, or will become, obsolete. Instead, employment services could consider the skills a truck driver might possess – logistics, knowledge of transport guidelines, communicating with dispatch or time management – that could help them succeed in a new role.

FROM JOBS FOR LIFE TO LIFELONG LEARNING

The key to having transferrable skills is lifelong learning and regular retraining – not as triage following the loss of a job but as a central and habitual part of participating in the workforce. The old 'career ladder' no longer exists. A better analogy would be, as American businessman Peter Guber puts it, a 'career triangle', where opportunities start at a point and broaden out.[24] People's work journeys are now 'nonlinear, seemingly circuitous, requiring varied skillsets and interdisciplinary talents'. Guber – who not only co-owns the Golden State Warriors NBA team and the fabled Los Angeles Dodgers baseball team, but is also involved in the music and digital media industries, among other pursuits – insists he would never have had the success he did if he considered his career to be a ladder with one solitary, pre-determined path upwards.

Our approach recognises that whether it be school, university, TAFE or an apprenticeship immediately after school, the education you have received by your early to mid-twenties will be insufficient to propel you for forty or fifty years of reliable work. Those days are over.

Our earlier recommendations about computational thinking in schools will help. As will a sixth policy idea: considering minimum STEM standards for university entrants and a compulsory top-up computational thinking course, even in non-STEM degrees. But a commitment to education must go deeper and further than one post-school qualification. Even if students receive the best possible education before they enter the workforce, their skills and knowledge have an expiry date – a problem that will grow with further advances in technology.

For continuous learning to be worthwhile for workers and employers, an immediate, tangible return from the training is needed. Standardised computational thinking or data analysis courses might not achieve this. More granular courses, such as an introduction to computational thinking for accountants, or data analysis for logistics workers, would have a greater direct pay-off. These courses could teach general principles in an applied way and highlight the way the skills can be used in specific fields.

Experience from mass redundancy scenarios shows that it's difficult to get workers motivated and engaged in training for entirely new careers – even when there's a known end date for an existing job. It's even harder when you're dealing with an abstract, future need. But many workers take pride in their current jobs and will invest time in getting better at them. In this way, the delivery of new skills like computational thinking and data analysis needs to look more like an apprenticeship, or a routine part of the working week, than a degree.

Australia could consider a version of Singapore's SkillsFuture initiative as a potential framework – our seventh recommendation. The beauty of this model is that it requires input from employers, who outline the expected changes in their industries over the next three to five years and list the skills they think workers will need to cope with, and thrive under, those changes.[25] Workers then learn those skills while they are still employed. They do not simply wait until they are excluded from work due to having obsolete skills. When announcing the scheme in 2015, Singapore's deputy prime minister, Tharman Shanmugaratnam, described 'a meritocracy of skills, not a hierarchy of grades earned early in life'.[26] The initiative heavily subsidises employers who train staff and offers S$500 credits for Singaporeans aged twenty-five and over to help them with out-of-pocket expenses for 'work skills–related courses'. This is a costly concept but one worth investigating.

Another expensive element of lifelong learning and retraining is the cost in lost wages or social security payments when Australians leave the workforce, either because they've lost their job or they've taken a break to study. A future Australian government will need to consider how to maintain living standards during these periods, or else the prospect of earning less for a long period may act as a deterrent.

For those who need to update their skills or learn new ones, the government could consider an eighth policy direction: a form of income-contingent loan to help pay for study or training, similar to HECS-HELP for university students. These loans would be repaid through the tax system once a certain level of income was reached, and repayments would increase as the level of remuneration increased. Income-contingent loans (ICLs) will serve as an anchor to ensure workers have enough support while they attempt to re-enter the workforce and do not become discouraged and drop out of the system completely. Any new proposal along these lines would need to learn the policy design lessons of the debacle with private training colleges, which has done such damage to the ICL concept by exploiting students and selling worthless qualifications.

This proposal could be extended to incorporate a ninth policy suggestion: learning accounts. These could provide tax advantages or even be matched proportionately by government, so that workers have a pool of funds to draw upon for training costs and wage subsidies. Australians could elect to pay a certain percentage of their salary into a learning account, or even continue to pay the same amount as their HECS-HELP loan after it has been fully repaid, so that they have a positive balance. A further step would be employer contributions to the accounts. Of course, there would be real concerns with possible exploitation, so a future government would need to develop a system that encouraged new, innovative and flexible providers, but also provided protection from the kind of rorting we have seen in recent times.

WORK AND WAGES

Steps to improve the social safety net and to encourage and fund lifelong learning are critical to advancing the fair go in the new machine age. Caring about inequality means caring about wages and working conditions too. Caring about wages and working conditions means caring about power relationships in the workplace, which are at risk of tilting further in the direction of the owners of the machines – the employers – as automation, machine learning, artificial intelligence and robotics replace more and more humans. Redundancies, less secure work, increasing casualisation, and worker exploitation are the emerging challenges. A job is one thing; ensuring work is properly rewarded is another.

For all these reasons, Harvard Law School professor Yochai Benkler's concept of 'counterpower' is a fascinating one. It refers to the responsibility of policymakers and decision-makers to recalibrate the power balance between the rights of workers and those of their bosses. He points out that 'power is everywhere, in markets as in states. Only building counterpower, political, legal, social, and technical, will lead to a more egalitarian distribution of wealth and income.'[27]

Unions will have a more important role than ever to play in securing worker protections in the new machine age, but face challenges in organising increasingly casual and mobile workforces. This is one reason why the conservative side of politics cynically supports casualisation under the guise of flexibility; it weakens unions and empowers employers. It's also why the ACTU, the Transport Workers Union and many others are all putting so much thought into worker protections in the new economy, as well as ways to encourage the use of contractors who meet minimum standards of pay, and to ensure Australians are saving superannuation for their retirement.

Along with unions, the other important part of the equation is robust legislated worker protections across the board, and especially in newer frontiers such as the gig economy,[28] to ensure workers don't fall through the cracks or miss out on protections, retirement incomes and entitlements they need and deserve.

American entrepreneur Nick Hanauer and labour union leader David Rolf have painted a bleak picture of how the sharing economy is eroding job certainty. Working several casual jobs instead of one full-time job and being hired for very specific tasks, they say, is undermining the middle class in America:

> A nation of independent contractors is a nation of workers without any of the benefits that defined the decent and dignified life that gave one reason to be optimistic about the future – a gross violation of the social contract that helped create the greatest economic expansion, the most dramatic increase in living standards, and the largest, most prosperous, most productive, and most secure middle class in human history.[29]

Hanauer and Rolf propose a 'shared security system' for America, where workers accrue benefits often paid by a full-time job, such as sick leave, health insurance, employment insurance, annual leave and superannuation. These benefits could be held in a centralised account, either by state or federal governments, and would be fully portable between jobs within the flexible environment of the sharing economy. Another version of this would be to transfer responsibility for payroll administration from employer to employee.[30] This would allow workers to outsource the management of entitlements to third party experts or advocates. Unions would be well placed to play this role, and it would open up an entirely new terrain for bargaining. Most importantly, it would mean entitlements weren't tied to the employer, as in the 'shared security system' example.

As workers change jobs more frequently, portability of entitlements and protections will be key. So the tenth policy direction is to ensure workers can take their accrued entitlements with them when they move from job to job, or when they are working multiple insecure jobs at once; that companies make pro-rata contributions; and that the genuinely self-employed save for retirement and insure against sickness or other risks. This could also address some of the problems we are already seeing in the gig economy.

Central to addressing those problems will be ensuring businesses in the gig economy don't undercut the wages and conditions of Australian workers. Our eleventh policy recommendation involves addressing those elements that threaten to exacerbate and accelerate inequality, by encouraging collective bargaining.[31] This policy direction is particularly important, given workers in the gig economy aren't considered to be employees and are not a homogenous group. Usually they can't negotiate prices or conditions on individual jobs, though they can choose not to do the job if the remuneration is insufficient. These issues will become more prominent in the wider workforce as a result of technological change, and also as client-directed service work – such as that associated with the National Disability Insurance Scheme – becomes an increasingly important source of new employment.

Another way to restore the power balance between employer and employee would be to ensure workers share in the gains of the business they work for. The author of *The Sharing Economy*, Arun Sundararajan, argues for a decentralised form of capital – crowd-based capitalism.[32]

Benkler, who we mentioned earlier, champions the idea of 'platform cooperativism', which he describes as 'the idea that the people who work and use networked platforms should own them'.[33] It's a term coined by New York academic Trebor Scholz, who argued that 'there isn't just one, inevitable future of work', and called for worker-owned cooperatives to cut out the middlemen and design their own app-based platforms to rival the likes of Uber.[34] Considering ways to encourage platform cooperativism to ensure workers have more say in the sharing economy is our twelfth policy recommendation.

This builds on an existing but under-utilised concept of employees as shareholders of the companies they work for. There is already a big push in the start-up sector for workers to be offered share options or employee share schemes as a way for companies to remain internationally competitive and attract high-level talent without paying high salaries.[35] Our thirteenth policy recommendation is to use employee shares not just as bait to attract talent, but as an anchor to ensure workers have a stake and a say in the company they work for. As technological change sees owners enjoy increased productivity gains and profits, employee shares would ensure workers get a slice of that success. Of course, care would need to be taken to ensure share offers weren't used as an excuse to cut take-home pay.

Workers' stakes in the companies that employ them will become increasingly important to keep inequality in check. That's why we would consider, as a fourteenth policy proposal, encouraging if not mandating worker representation on boards. British prime minister Theresa May championed this idea when she sought the Conservative leadership, taking inspiration from a longstanding German model.[36] In Germany it is called 'codetermination', and it mandates worker representation on boards according to the size of the company.

TAX THE ROBOTS?

One of the most radical policy suggestions has come from Nobel laureate and leading economist Robert Shiller, who proposes that governments should levy taxes on the robots themselves.[37] Shiller's position is that even a temporary tax, targeted at robots rather than directly imposed on higher-income earners, would be a palatable way to help address rising inequality. Better yet, the revenue generated could be spent on retraining displaced workers, again helping to bridge the divide. As Shiller puts it, 'this would accord with our natural sense of justice, and thus be likely to endure'.

Shiller is not alone in calling on governments to 'tax the robots'. The United Nations Committee on Trade and Development has made similar calls, as has billionaire Microsoft founder Bill Gates, who argued that if a robot replaces someone who does $50,000 worth of work in a factory, then the robot should be taxed at the same level.[38] He concedes that working out how to apply the tax would be challenging, but insists the extra productivity robots will generate should result in more taxes to help fund measures to ameliorate inequality.[39]

The concept of taxing robots has plenty of critics with legitimate concerns, such as the difficulty in defining what a robot is and the risk of businesses moving offshore to countries without a similar tax. Larry Summers called it 'protectionism against progress' and argued a better response would be higher and more progressive income tax.[40]

There are broader considerations to factor in. For example, Obama's office, in its *Artificial Intelligence, Automation and the Economy* report, noted that capital income was often taxed at lower rates than labour income, or not at all.[41]

We are not ready to recommend a robot tax, but our fifteenth policy suggestion is for experts and policy-makers to turn their minds to whether, and how, a robot tax could work in Australia. This effort should take into account the mobility of capital and international developments. A policy of this kind is nowhere near ready to implement, but there is no downside to investigating and closely following its evolution elsewhere.

MEASURING AND MONETISING

The next three policy directions are intended to help measure and monetise the contribution humans make to the creation of value in the new machine age. The current suite of economic data, especially the gross domestic product headline numbers, inadequately measures a raft of important considerations, such as technology's contribution to increased productivity.

The use of GDP as a measure of a nation's success or its citizens' wellbeing has been criticised since at least 1968, when US presidential candidate Robert Kennedy said 'it measures everything, in short, except that which makes life worthwhile'. The problem in the digital age is that GDP doesn't capture the intricacies of the modern economy. More recent critics point out that GDP – developed during the industrial age – distorts the economy as it's skewed towards manufacturing and doesn't take into account digital services that aren't technically 'bought' or 'sold'. Nor does it sufficiently acknowledge the labour-saving effect of new technologies.[42]

Although GDP is increasingly criticised, there is no agreed replacement, nor a consensus around anything other than augmenting it with other measures. A minimal change would be to shift the emphasis to real net disposable income per capita. Some argue we should go further, preferring a genuine progress indicator, which takes into account factors such as income distribution, the value of household and volunteer work and the costs of crime and pollution.[43]

Governments need to agree to a new statistical basis that takes technology's contribution into account, and make those stats freely available – our sixteenth recommendation. This should be part of a broader effort to measure and publish data on inequality and social immobility.[44]

Another key reason to modernise our statistical base is to help find new ways to monetise people's contribution to companies that accrue massive profits, while those who actually provide them with their 'free' products, like metadata or information, receive targeted advertising and marketing but no financial compensation. Our seventeenth recommendation is to explore feasible ways to design a market for individuals' information.

Our eighteenth recommendation recognises the difficulties in compensating consumers for the information they provide and focuses on using that information for the public good. That could involve removing barriers for researchers wanting to model how changes to social security payments will impact our most vulnerable, or even how health and education funding should be spent.

A well-formulated national information policy would allow governments to provide researchers with access to private data, as long as it was for the public good. An independent data council could identify the necessary legislative and regulatory changes to ensure the data provided was useful, but still confidential and de-identified to address privacy concerns.[45]

The council could also complete a comprehensive audit of government data sets and develop a personal information management system. It's pleasing to see the Productivity Commission has picked up on this agenda.[46] A targeted policy that optimises the generation, protection, access and use of information in Australia would add to our research capacity and attract top-level start-up investment and talent.[47]

Our nineteenth recommendation is already relevant today, but will become even more so: the need for universal access to high-speed, ubiquitous and affordable broadband. It is increasingly essential for Australians to be able to access commercial services such as e-banking and e-shopping; high-quality healthcare, including remote healthcare monitoring; online government services; remote education and training; and entertainment. It is also needed to support the growth of home working and decentralisation. Allowing only a portion of society to access the benefits of fast broadband will turbocharge inequality.

COLLABORATION

Underpinning all of these recommendations is the recognition that, to date, policymakers have mostly failed to keep pace with technological change. The obvious problem, as spelled out by the Brookings Institution, is that 'major societal changes occur without a public dialogue or meaningful government involvement – at least not until after the "facts on the ground" are already established'.[48] With that in mind, our twentieth recommendation picks up the Brookings recommendation for public–private partnerships, made up of government, businesses and unions, to establish policy frameworks and regulations before they are required – not after.

There are countless other ways to meet the objectives of inclusive economic growth, reward for work and a decent social safety net over a ten- to twenty-year timeframe. Governments should also deploy additional measures, especially in policy areas like tax and service delivery, which will address inequality in general.

The social wage, including decent healthcare, superannuation and progressive taxes are important bulwarks.

In addition to a world-class education and training system, they go some of the way – though not all of the way – to ensuring our country is one worth basing businesses in, relying on superior workforces and systems. But our specific concern with technology and inequality in the labour market has limited us to twenty broad domestic policy directions to go with the seven in the previous chapter and eight to come in the next.

In summary, our recommendations in this chapter are:

1. A 'social safety web' that uses big data for good in the social security system.
2. A focus on 'caring services' skills, augmented caring roles and additional work opportunities for people with a disability.
3. A rewrite of reciprocal obligation.
4. Income smoothing.
5. Labour market programs that match skills, not jobs.
6. Minimum STEM requirements for university and compulsory computational thinking courses.
7. Lifelong learning programs, like Singapore's SkillsFuture initiative.
8. Better income-contingent loans to help workers retrain.
9. Lifelong learning accounts.
10. Portable entitlements.

11. Encouragement of collective bargaining in the sharing economy.
12. Development of 'platform cooperativism'.
13. Employee shares.
14. 'Codetermination' and worker representation on boards.
15. Careful consideration of a robot tax.
16. Augmentation or replacement of GDP as the headline economic measure, and regular publication of data on inequality and immobility, causes and consequences.
17. Exploration of markets for information.
18. Use of private data for public good.
19. Universal access to fast and reliable broadband.
20. Collaborative public–private approaches.

If a future government considers each of these points, Australia will be well placed to reap the benefits of technological change and well fortified against the potential transitional pitfalls that could exacerbate and accelerate inequality.

But no government can tackle this challenge alone, solely within its own borders or without the engagement of citizens. We also require a change of mind at the grassroots; a culture and love of learning and self-improvement; and a communal appreciation and understanding that we need to reposition ourselves if we are to survive and thrive in the labour market of the new machine age. That's why the next chapter touches on what individuals will need to consider in their own lives.

CHAPTER 7
WHAT EACH OF US CAN DO

Some decades ago Mike spent time in Shanghai working on the transfer of technology from Australia to China. The deal was for telecommunications equipment designed and manufactured in Australia to be packaged into a final product and used as part of China's rapidly expanding telecoms infrastructure.

The product was designed in an Australian development lab, and all the electronics and associated software were built in a modern Australian factory and shipped to China. The metalwork that housed the electronics was based on an Australian design but made in China. From the time of those early visits, and through numerous subsequent visits to Shanghai, it was never in doubt that the management of the Shanghai operations had a long-term plan to design the hardware, write the software and build the entire product, or the next generation of the product, in China. On the first visit the Chinese management provided a tour of the facilities they were building to accommodate the team of engineers they had already employed, including several engineers with PhDs in telecommunications.

You probably know the end of this story. A quick glance at the Wikipedia page on 'Telecommunications equipment' shows which country now leads the world in the design and manufacture of this type of sophisticated equipment.

That the Chinese had a plan to become the leader in telecommunications products would be no surprise to anyone. But the speed and determination with which they executed this plan was quite remarkable. They not only made large investments in infrastructure and physical facilities but also in education and training. There was a palpable ambition to succeed and an enthusiasm to learn that was shared by the management and workers of the Chinese company – supported, of course, by the relevant Chinese government ministries.

At the time this was happening it was common to hear in Australia that if we could not be competitive then we should exit local design and manufacture. That was what the large North American and European telco equipment companies ultimately did in Australia. They progressively transferred their Australian design and manufacturing facilities to other parts of the world, leaving their Australian operations to manage sales and service. Many of these same large North American and European companies have either ceased to exist or have been forced to merge to remain competitive and to move much of their manufacturing to cheaper locations.

For some time it has been conventional wisdom that Australia will import sophisticated manufactured goods, especially technological products. With the growing number of machines interconnected via the internet and the increasing sophistication and versatility of 3D printers, however, the trend of moving manufacturing to countries with low labour costs could potentially be reversed. Manufacturing will be able to target personal preferences more and more specifically. This presents Australia with an opportunity to rebuild its expertise in manufacturing – but with quite a different model to that of the past. It would be a good outcome for Australia if sufficient expertise now resides in Australia for this to become a reality.

This familiar story reminds us why we should ensure we have the type of workforce with the skills and expertise that can support businesses that want to invest in robots and automation on our shores rather than overseas. Such businesses will also be influenced in their decisions by the availability of economic (including technological) and social infrastructure.

With the view prevailing that Australian manufacturing was uncompetitive, the services sector has come into focus as the area in which Australia can compete on a world stage. This means providing, especially into the Asian market, educational services, project management, financial and agricultural consulting and their technology offshoots – FinTech and AgTech, for example. Services now represent some 70 per cent of the Australian economy – but only about 20 per cent of our exports.

China provides enormous scope for export of services. In 2016 China was home to approximately 240 million young people for Australian educators, 150 million cars on the road for Australian insurance companies and 80 million people aged over seventy for Australian aged-carers. This begs the question: who will be best equipped to address that market?

We aren't yet prepared to engage with China – and other countries around Asia – in a way that makes the most of these opportunities.[1] A focus on education and training, particularly in STEM disciplines, is a priority for many of our Asian neighbours. India has already demonstrated its ability to compete for service jobs. Australia may currently be happy to outsource what are regarded as low-level service tasks, since we are basing our expectations on more sophisticated service jobs. But we would be naive to assume that India will remain content to provide only the lower-skilled service jobs. Indian companies already fill thousands of IT roles in the US market, both remotely and as residents.

Engineers and technologists form a much higher percentage of university graduates in India and China than in Australia, and the number of technology graduates in some of our developing neighbours has been steadily increasing.[2] This puts Australia at a relative disadvantage in the competition to provide technology-based or technology-supported jobs. Despite having an education system that is available to more of the population for a greater period of their lives, compared with key economies in the Asian region, Australia is falling behind in mathematics and science competency.

Many future jobs, including service jobs, are going to be complemented and supported by technology and increasingly sophisticated computational tools. Countries that are equipping their people with a sound understanding of scientific and mathematical reasoning, competence in the use of technology, critical thinking skills and data analysis skills are going to capture the bulk of these jobs. We may not know exactly what those jobs will be, but we can make a pretty good guess at the skills and competencies likely to be needed.

Without a coherent strategy to address these issues we should not be surprised if the same countries that pushed aside many of the world-leading companies in high-technology product design and manufacture do the same thing with services.

To recap some key reasons we are in this comparatively weaker position:

- Insufficient numbers of experienced and skilled mathematics and science teachers in high schools.

- The lack of an incentive for students to undertake the more challenging levels of mathematics and science subjects, and a lowering of university prerequisites for mathematics and science subjects for some courses. Hopefully, both of these problems are now beginning to be reversed throughout Australia.
- A common view that it is not a major problem to be relatively uneducated in mathematics, science and technology.
- The lack of appreciation, by many Australian businesses, of the value of employing highly educated research personnel. Australia's chief scientist, Dr Alan Finkel, recently noted that only about one-third of PhD-trained researchers were employed in the business sector in Australia, compared with two-thirds in the United States.[3] This impedes research collaboration, which in turn impacts innovation.

None of the above means that Australians aren't avid users of technology. We clearly are; in general, Australians are early and rapid adopters of new consumer technologies. But this technology, on the whole, is designed and built in other countries. The loss of skills in designing and manufacturing products, particularly technology products, is not generally seen as something we should be concerned about. While employment in manufacturing has fallen, employment overall has risen. But there are some important areas of expertise that can be very effectively learned in a manufacturing environment, with its short cycle times and directly measurable performance parameters. These include disciplines such as scheduling, inventory management, process improvement, operations research and statistical process control. Knowledge in these areas, and with the technology-based tools that will support them, is likely to be increasingly important in the optimisation of service businesses in both domestic and export markets.

In recognising the problems with skill deficiencies mentioned above, the principal aim of this final chapter is to suggest actions an individual might take to prepare for a time when employment is heavily affected by technology, particularly robotics and AI. In making these suggestions we recognise that schools and governments can only do part of the job; a change of personal mindset is also necessary.

We know we cannot and should not prevent technology's march, so how do we change a national mindset? Through self-education; mentoring; reconsidering community attitudes to maths and science; taking steps to increase the number of girls taking STEM subjects; emphasising decision-making skills and pattern recognition; using computational thinking apps, learning tools and games; making sure our early education and intervention programs include technological skills; and building communication skills associated with drama, debating and advocacy.

SELF-EDUCATION

Whether you are a student, a parent or guardian, an employee confident of keeping your current job or one concerned your role may be displaced or significantly changed by technology, there is little to be lost by increasing your knowledge and skills in the areas that are likely to be important. It is not as difficult as it may sound to become better informed about mathematics, science and computational thinking.

There is an almost unlimited amount of educational material available online from sites such as the Khan Academy, iTunesU and YouTube. Most universities, including Harvard, Stanford and Yale in the United States, run massive open online courses (MOOCs) outside the normal education system. All of this is making education more accessible.

But it is usually people who have already undertaken some type of higher education who sign up for such courses. The challenge for a future government is to encourage lifelong learning and provide reskilling options for those who are not already well educated; who are content with an existing low- or middle-paid job; and who wouldn't reskill or embark on further study if it wasn't necessary. As an article in *The Economist* put it, 'There is no natural pathway from trucker to coder.'[4] But while the path may not be natural or easy, it is doable. Each one of us can encourage and assist a relative or someone we know to begin this journey using the educational assets already available to everyone.

But we should be cognisant of the challenge faced by someone who has been out of the education system for many years and is now being encouraged to study unfamiliar subjects. Even if well motivated it is still quite a challenge – not unlike the prospect of learning a new language later in life.

This is why the Singaporean SkillsFuture program mentioned in the previous chapter is likely to be very valuable. Introductory courses in mathematics, statistics, science and computational thinking could be developed that are linked to specific industry areas such as building and construction, transport or retail. These courses could be made available to people in work and those seeking employment, delivered as part of the existing TAFE system or similar.

Taking an even more gradual or staged approach, there are many excellent books about mathematics and science that can make a good introduction to these subjects. Many of these books do not require prior mathematical or scientific knowledge and provide the layreader with an interesting perspective on the subject, often along with historical insights. Bill Bryson's *A Short History of Nearly Everything* is an excellent, entertaining example, as is *Surely You're Joking, Mr. Feynman!*, a short memoir written by the Nobel Prize–winner Richard Feynman. Feynman was both a great physicist and a great educator and he demonstrates that even physicists can have a sense of humour!

Learning how to avoid, or at least recognise, biases is also worth the investment of some time. One of the half-dozen skills frequently mentioned as important for future jobs is decision-making. And good decision-making depends on a number of other abilities, including avoiding falling prey to our own or other people's biases.

Two excellent books about the various types of biases and how we can categorise and recognise them are Nobel laureate Daniel Kahneman's 2011 book *Thinking Fast and Slow*, and psychology professor David Myers' 2002 book *Intuition: Its Powers and Perils*. Reading these two books is an excellent way to become aware of how biases impact our decision-making. This is not only of value at work; we also need to make good decisions in our daily lives.

CHANGING HOW WE THINK ABOUT MATHS AND SCIENCE

When it comes to maths and science, we have a cultural problem. Most people would find it embarrassing to say 'I am hopeless at reading and writing' but it is quite acceptable, in our culture, to say – and often with pride – 'I am hopeless at maths'. If we want Australian youth to grow up without a bias against mathematics and science, and instead to realise that they are an indispensable part of modern life, these subjects need to become a part of their early life, just as reading and writing do today. Most young children are naturally interested in numbers, shapes and measurements.

We would be wise to prepare our young people with the skills to prosper in the world of their future. We will be doing them a disservice if we, through our words and actions, convey apathy about or even disdain for maths and science. Not only is a lack of understanding of these areas a problem for living in a technology-dominated world, it is also a great pity; people unfamiliar with mathematics and science are missing out. One does not have to be a highly proficient artist, writer, actor, poet, dancer or composer to appreciate the wonder of paintings, literature, drama, poetry, dance or music. Likewise, one does not have to be a proficient scientist or mathematician to appreciate the beauty inherent in science and mathematics. They are not purely utilitarian tools but rather ways of revealing the underlying structure and elegance of the universe – both animate and inanimate. Failing to equip our young people with the ability to see this additional dimension of the world around them is potentially to limit their appreciation of the world.

There is an important practical element as well. Fostering an understanding of and appreciation for maths and computational thinking will help today's students be more comfortable working in tomorrow's jobs. We want those who will be interacting with machines and intelligent systems to have more than just a passing familiarity with them, particularly in situations where lives are at stake, as is the case with surgeons or even pilots.

So from many perspectives, it is vitally important that young people see mathematics and science as important subjects to master. Each of us, as parent, grandparent, sibling, teacher or friend, needs to help break the cycle that leaves many of Australia's youth with the view that showing a passion for, or even an interest in, maths or science is peculiar or 'nerdy'. We can do this through the language we use with young people, the interest we show in what they are studying, the help we offer and the games we play with them. As we all know, children are very perceptive in seeing the differences between what we say and what we do. So if children experience a close adult actually reading, studying or discussing these subjects regularly, it will make a difference to the attitudes they develop.

GENDER AND TECHNOLOGY

Another important issue is the gender bias in mathematics, some of the sciences and technology. Globally, there continues to be relatively low numbers of women studying and working in science, technology, engineering and mathematics, despite the majority of students enrolled in higher education in Australia being female. Much has been written on this subject, and it will take a concerted effort by all of us to combat this issue.

At Mabel Park State High School in Logan, Queensland, dedicated staff have begun a new program called GEMS, or Girls Excelling in Maths and Science. Already they have succeeded in attracting young women to STEM subjects, supporting the students who participate with extra activities and help, bringing in inspiring mentors, building partnerships with local primary schools and universities simultaneously, and more. GEMS has already won a number of awards, and other schools are looking to replicate this model.

Several organisations are taking valuable initiatives, including a collaboration between the Australian Mathematical Sciences Institute and the BHP Foundation called CHOOSEMATHS, which aims to turn around community attitudes to mathematics, especially among girls and young women. No doubt BHP can see the benefits of having more highly trained workers available for its business.

A 2015 research paper based on a long-term study in Israeli schools estimates the effect of the gender bias on the part of primary school teachers of mathematics and science.[5] The average marks given by teachers who did not know the gender of the students they were assessing were compared with the marks given by teachers who did know the students' gender.

The report concludes that the teachers tend to 'over-assess' boys in primary school. To put it bluntly, student assessment is biased in favour of boys. This has a positive effect on boys' achievement and also on their successful completion of advanced courses in mathematics and science in high school, and a negative effect on girls.

Addressing this type of problem is not easy because these biases are usually unconscious. And we would be naive to imagine that we, as individuals, are free of such biases. This means we all need to be vigilant not to assume, when interacting with young people, that there is any inherent difference in the capabilities of girls and boys. This is especially the case for discussions, play and expectations around mathematics, science and technology. Given our unconscious biases, it would be even better to plan frequent interactions with girls on these subjects.

STATISTICS AND PATTERNS

Statistical thinking is an important related ability. Like computational thinking it is underpinned by mathematical concepts, but it also relates to avoiding biases. It is not uncommon – in business or in many other circumstances – for people to jump to conclusions about trends or patterns that are not supported by the data. This is a natural human propensity, possibly a consequence of our evolution as pattern-recognising creatures.

So, the ability to recognise patterns, to know when there really is evidence of a trend, to understand the differences between causation and correlation and to be aware that there is natural variation in all processes is a very valuable set of skills. While overreacting to natural or random variation can be a problem, it can also be risky to ignore a real trend with strong statistical evidence – climate change being an obvious example.

But statistics is not an area with which most people are familiar or feel comfortable. It suffers from the same image problem as mathematics, appearing to be dependent on understanding complex equations and abstruse concepts. While it is true that some statistical problems require sophisticated mathematics, it is possible to understand many important statistical concepts and develop an ability to think statistically without mastering advanced mathematics.

Understanding basic statistical concepts is valuable in almost every field – health, agriculture, manufacturing, finance, transport, communications and many others – so it has enormous potential to improve products and processes in almost all organisations. It is also useful in daily life for understanding the results of health checks, the reality of gambling risks, what advertisers are saying and the findings of social surveys. And as we have seen in earlier chapters, probabilistic and statistical techniques are involved in many of the machine-learning strategies encountered in AI and robotics.

Once again, there is a large amount of material available to the interested layperson that will provide a good overview. An excellent book that provides many examples of how statistical thinking can be applied in everyday life, and does so using only basic arithmetic, is Gerd Gigerenzer's 2002 book *Calculated Risks: How to Know When Numbers Deceive You*. For parents, grandparents or guardians of school-age children there is another motivation for learning some statistical concepts: probability and statistics are likely to be introduced into the maths curriculum. Being able to assist students will emphasise the importance of this subject and hopefully activate their interest in pursuing more advanced studies.

PROBLEM-SOLVING

In earlier chapters we highlighted the need to acquire computational thinking skills and suggested integrating computational thinking into maths and science subjects. There are two actions that individuals could take in this area. The first is to gain some practical familiarity with computational tools oneself, by learning the basics of a high-level computational language and seeing how such a tool can be applied to real problems. This would provide insight into some important elements of computational thinking. Once again, this is not as hard as it might at first appear. There are computational applications that are relatively inexpensive and some that are available online at no cost.[6]

The second is to sit down with a child – or indeed another adult – and work on a problem together using such tools. This may not at first have the same appeal as a day out at the cricket but it can be quite rewarding to jointly solve a problem. And helping to prepare ourselves, our children and our grandchildren for a world where technology plays a prominent role is certainly a good investment of time. Along similar lines, there are quite sophisticated robotics kits available, such as the Lego Mindstorms robotic kit, which aims to build critical thinking and computer science skills. While this kit is not inexpensive, it provides an opportunity to engage with and develop skills in one of the important future technologies.

Making an early start is especially valuable. Parents and guardians of preschool children can help foster a love of learning in their young ones through initiatives such as the HIPPY program, operated by the Brotherhood of St Laurence, with funding provided by the Department of Social Services. Organised on a local community basis, the program is free and raises the awareness of parents' role as their child's first teachers. Tutors visit families in their homes every two weeks to provide guidance and mentoring. The home tutors have been trained by a local coordinator and have completed the activities with their own children. HIPPY is a very valuable program, based on the evidence that encouraging numeracy and literacy in a child's early years makes a positive difference to their success in learning throughout the rest of their life.[7]

DEMAND FOR SCIENCE GRADUATES

It needs to be acknowledged that currently, while government, educators and the media are focused on STEM subjects, the demand for science graduates in Australia is not growing. A 2016 Grattan Institute report observes that the percentage of science graduates with a bachelor degree who have gained a full-time position four months after graduating is well below the average for all graduates, at only 51 per cent.[8] Engineering graduates fare considerably better.

While this appears to be a disturbing statistic, there may be a number of reasons, as the Grattan report mentions. There was a surge in domestic student enrolments in science starting in 2009, and large proportion of science bachelor degree graduates go on to do honours programs and postgraduate study. Others use a science degree as a stepping stone to study medicine. Given the likely growth in the importance of technology, it would be a mistake to conclude that today's poorer prospects for science graduates will continue into the future. But it may suggest a greater focus on strongly analytical science subjects such as mathematics, physics and statistics.

COMMUNICATION AND PERSUASION

We cannot, of course, forget that there are still human qualities vital to many jobs that no computer will be able to replicate in the near or mid-term future. So focusing on enhancing these qualities may also be a worthwhile investment. This idea was elegantly captured in a recent *Harvard Business Review* article:

> Those that want to stay relevant in their professions will need to focus on skills and capabilities that artificial intelligence has trouble replicating – understanding, motivating, and interacting with human beings. A smart machine might be able to diagnose an illness and even recommend treatment better than a doctor. It takes a person, however, to sit with a patient, understand their life situation (finances, family, quality of life, etc.), and help determine what treatment plan is optimal.
>
> It's these human capabilities that will become more and more prized over the next decade. Skills like persuasion, social understanding, and empathy are going to become differentiators as artificial intelligence and machine learning take over our other tasks.[9]

One very practical way to build skills in understanding and interacting with others is to take a course in drama. This is particularly valuable for young people who can be shy about talking publicly. If this is followed by some time in a debating team to sharpen critical thinking, communication skills and persuasion, an individual will have a fairly rounded set of experiences.

THE PATH AHEAD

Advancing the fair go in the new machine age will require not only a change of personal mindset, but changes to our schools and policy agendas as well. As citizens, students, governments – and, above all, as human beings – we can't muddle through what is to come: we must forge a deliberate and considered path ahead. We must rethink the way we teach our young people, the way we make transitions between jobs that turn over ever more rapidly, the power relationships at work, the way we monetise value and how we fairly and efficiently raise taxes.

We urge anyone concerned about the issues raised in this book to make your views known to your schools, community groups and political representatives. These are matters on which public debate is essential.

This book has suggested a large number of initiatives for consideration – thirty-three in all. Some are fully formed and could be enacted now, but most require substantial further thinking and time. In summary, they are:

SCHOOLS (CHAPTER 5)

1. Training and mentoring more STEM teachers.
2. More, and more substantial, 'flipped learning' trials.
3. Adoption of Queensland's 'digital solutions' curriculum.

4. Compulsory coding and robotics in primary and secondary schools.
5. Emphasis on computational thinking in existing subjects.
6. Needs-based funding for schools to combat technological inequality.
7. Early education and intervention, especially in poorer communities.

GOVERNMENT (CHAPTER 6)

1. A 'social safety web' that uses big data for good in the social security system.
2. A focus on 'caring services' skills, augmented caring roles and additional work opportunities for people with a disability.
3. A rewrite of reciprocal obligation.
4. Income smoothing.
5. Labour market programs that match skills, not jobs.
6. Minimum STEM requirements for university and compulsory computational thinking courses.
7. Lifelong learning programs, like Singapore's SkillsFuture initiative.
8. Better income-contingent loans to help workers retrain.
9. Lifelong learning accounts.
10. Portable entitlements.

11. Encouragement of collective bargaining in the sharing economy.
12. Development of 'platform cooperativism'.
13. Employee shares.
14. 'Codetermination' and worker representation on boards.
15. Careful consideration of a robot tax.
16. Augmentation or replacement of GDP as the headline economic measure, and regular publication of data on inequality and immobility, causes and consequences.
17. Exploration of markets for information.
18. Use of private data for public good.
19. Universal access to fast and reliable broadband.
20. Collaborative public–private approaches.

INDIVIDUALS (CHAPTER 7)

1. Self-education – online courses, computational thinking apps, reading.
2. Family, school and community mentoring.
3. Change in community attitudes towards science and maths.
4. Extra attention and programs for girls.
5. Emphasis on statistical thinking.
6. Building communication skills associated with debating, drama and advocacy.

There will be countless other ideas than those sketched out here, and they are welcome even when they differ from ours. But let's not use the unpredictability of the technology boom's trajectory as an excuse to do nothing. Or, just as unhelpful, let's not try to retrofit the policies of the past, or limit our ambition to merely attempting to triage for those left behind by change.

Equally, let's not pretend we can hold these changes back, or that we can let them rip without caring about how the effects – substantially good but potentially bad – are distributed in our society. Let's not obsess over refereeing competing claims about jobs lost, created, changed or augmented. We don't know in what numbers, but we do know that some jobs will be eliminated, some jobs will be created, some jobs will be augmented and all jobs will change. That's more than enough for us to know that we need to change too.

We need to agree that as automation, robotics, artificial intelligence and machine learning transform our workplaces, we have choices. We can decide whether or not bursts of technology need also be accompanied by bursts of inequality or immobility.

This book's focus on technology and jobs is all about helping Australians and their representatives understand the choices they face in the new machine age. We celebrate technology's potential to improve lives, save time and effort, and boost productivity. But we also acknowledge that, left unattended, technology could turbocharge already worrying trends towards inequality. People are already anxious about what the future will hold, where jobs will come from and what their kids will do.

Above all, this book is a call to reject technological trickle-down, which assumes that the economic gains achieved by artificial intelligence, automation, machine learning and robotics will flow freely to all without leadership, collaboration and foresight. Even when machines bring us new sources of prosperity and wealth, there is no guarantee that we will all rise together. Australians still cherish the fair go, but cherishing it will not be enough to ensure it is a defining feature of our future and not just our past. To do that, we have to care enough to act.

ACKNOWLEDGMENTS

The most enjoyable aspect of writing *Changing Jobs* has been the opportunity to collaborate with each other and with so many friends who understand technology and care about the future of our country.

We wouldn't have completed it without Nathan Paull's months of sustained and dedicated effort. We thank him, as well as Elliot Stein and Michael Quinlivan, Mitchell Watt and Emily Millane – all of Jim's current and former team – as well as the parliamentary library for their assistance with facts and figures.

We wouldn't have started down this path without the intellectual infrastructure provided by Andrew Charlton and the 'Courtyard Group' of economists, academics, analysts and experts who assemble periodically at his Sydney AlphaBeta office and at the University of Melbourne. We are grateful to Andrew, and to everyone who has participated in the group's discussions and debates about this book as it evolved from Mike's initial presentation, and especially those, such as Deborah Cobb-Clark and others, who provided very detailed feedback.

We wouldn't have contemplated this book without the encouragement and faith of Chris Feik, our publisher at Black Inc. It would be twice as long and half as good without the careful editing of Jo Rosenberg. We thank them, and Head of Marketing and Publicity Kate Nash, for their professionalism, dedication and enthusiasm.

Some very busy family and friends selflessly pored over draft chapters, including Rod Tucker, Scott Nelson, Vincent Quigley, Mike Kaiser, Chris Barrett, Michael Cooney, Amit Singh, Annie O'Rourke and Laura Chalmers.

We acknowledge that many of Jim's parliamentary colleagues are working on some or all of the issues covered by this book, including Wayne Swan, Brendan O'Connor, Kim Carr, Ed Husic, Michelle Rowland, Kate Ellis and a long list of others. We thank those who discussed the project with us and provided detailed advice, including Chris Bowen, Tanya Plibersek and Jenny Macklin, who engaged on specific policy ideas, and Tim Watts and Andrew Leigh, who gave considered feedback on the entire manuscript.

We are also very grateful to Bill Shorten for fostering a culture of ideas and fresh thinking in the Labor Party, which encourages a project like this one.

Every cent of the authors' royalties from the sale of the book will go to four schools in Jim's electorate: the robotics program at Browns Plains State School; the learning centre at Woodridge North; the Girls Excelling in Maths and Science (GEMS) program at Mabel Park High; and Code Club at Springwood Central Primary. Buying this book supports students and teachers who are doing their best, as we all should, to prepare for, and succeed together in, the new machine age.

Jim Chalmers and Mike Quigley
Logan and Sydney
August 2017

NOTES

Preface

1 John Kenneth Galbraith, 1977, *The Age of Uncertainty: A History of Economic Ideas and Their Consequences*, Houghton Mifflin, Boston, p. 330.
2 Michael Ignatieff, 2014, 'We Need a New Bismarck to Tame the Machines', *Financial Times*, 11 February.

Chapter 1

1 Aaron Smith and Janna Anderson, 2014, 'AI, Robotics and the Future of Jobs', Pew Research Center report, http://www.pewinternet.org/2014/08/06/future-of-jobs/, 6 August.
2 Erik Brynjolfsen and Andrew McAfee, 2014, *The Second Machine Age: Work, Progress, and Prosperity in a Time of Brilliant Machines*, W.W. Norton & Co, New York, p. 47.
3 James Reinders, 2011, Intel, *Rich Report*, www.youtube.com/watch?v=I4tkZVe2f2A, 22 November.
4 Nick Bostrom, 2014, *Superintelligence: Paths, Dangers, Strategies*, Oxford University Press, Oxford, pp. 19–21.
5 Demis Hassabis and David Silver, 2017, 'AlphaGo's Next Move', DeepMind, https://deepmind.com/blog/alphagos-next-move, 27 May.
6 Mark White, 2016, '"I Wasn't Interested in Just Following the Rules": Data Scientist Jeremy Howard and the "Next Internet"', *Sydney Moring Herald*, www.smh.com.au/technology/innovation/i-wasnt-interested-in-just-following-the-rules-data-scientist-jeremy-howard-and-the-next-internet-20160419-go9rps.html, 29 May.

7 Andrew Keen, 2015, 'Ascent of the Robots Has Begun, and It's Not Good for Humankind', *The Australian*, 27 February.

8 Carl Benedikt Frey and Michael A. Osborne, 2013, 'The Future of Employment: How Susceptible are Jobs to Computerisation?', www.oxfordmartin.ox.ac.uk/downloads/academic/The_Future_of_Employment.pdf, 17 September.

9 Mark Reading, Jeremy Thorpe and Tony Peake, 2015, 'A Smart Move', PWC report, www.pwc.com.au/pdf/a-smart-move-pwc-stem-report-april-2015.pdf; 'Australia's Future Workforce?', Committee for Economic Development of Australia, June, http://adminpanel.ceda.com.au/FOLDERS/Service/Files/Documents/26792~Futureworkforce_June2015.pdf.

10 Carl Benedikt Frey, 2015, 'The End of Economic Growth', *Scientific American*, January, p. 12.

11 Jaron Lanier, 2013, *Who Owns the Future?*, Penguin Books, London, p. xx.

12 Brynjolfsen and McAfee, 2014, *The Second Machine Age*, p. 129.

13 OECD, 2017, 'OECD Economic Surveys: Australia 2017', www.oecd.org/eco/surveys/economic-survey-australia.htm, March, p. 2.

14 'The Mathematical Sciences in Australia: A Vision for 2025', Australian Academy of Science, Canberra, www.science.org.au/files/userfiles/support/reports-and-plans/2016/mathematics-decade-plan-2016-vision-for-2025.pdf, March 2016. Graph from Michael Evans and Frank Barrington, Year 12 Mathematics Participation Rates in Australia 1996–2015, data collection commissioned by AMSI.

15 'Australian and New Zealand Consensus Statement on the Health Benefits of Work', The Royal Australasian College of Physicians, 2011, www.racp.edu.au/docs/default-source/advocacy-library/realising-the-health-benefits-of-work.pdf.

Chapter 2

1 Yuval Noah Harari, 2015, *Sapiens: A Brief History of Humankind*, Vintage Books, London, pp. 89–90.

2 Fernand Braudel, 1981, *The Structures of Everyday Life: Civilisation and Capitalism 15th–18th Century, Volume 1*, Harper and Row, New York, p. 385.

3 Richard Feynman, 1964, speech, www.youtube.com/watch?v=0KmimDq4cSU.

4 See David Wootton, 2015, *The Invention of Science: A New History of the Scientific Revolution*, Allen Lane, London, p. 54; Stephen Weinberg, 2015, *To Explain the World: The Discovery of Modern Science*, Allen Lane, London, p. 147; and Moti Ben-Ari, 2005, *Just a Theory: Explaining the Nature of Science*, Prometheus Books, New York, p. 185.

5 See James Ang, Rajabrata Banerjee and Jakob Madsen, 2013, 'Innovation and Productivity Advances in British Agriculture: 1620–1850', *Southern Economic Journal*, Vol. 80, Issue 1.

6 Joel Mokyr, Chris Vickers and Nicolas L. Ziebarth, 2015, 'The History of Technological Anxiety and the Future of Economic Growth: Is This Time Different?', *Journal of Economic Perspectives*, Vol. 29, No. 3, www.aeaweb.org/articles?id=10.1257/jep.29.3.31.

7 A.M. Turing, 1936, 'On Computable Numbers, with an Application to the Entscheidungsproblem', *Proceedings of the London Mathematical Society*, Vol. 42, No. 1, www.dna.caltech.edu/courses/cs129/caltech_restricted/Turing_1936_IBID.pdf.

8 Cristopher Moore and Stephan Mertens, 2011, *The Nature of Computation*, Oxford University Press, New York, p. 297.

9 Andrew Hodges, 1983, *Alan Turing: The Enigma*, Vintage Books, London, pp. xvi–xvii.

10 Claude Shannon, 1948, 'A Mathematical Theory of Communications', *Bell System Technical Journal*, Vol. 27, http://math.harvard.edu/~ctm/home/text/others/shannon/entropy/entropy.pdf.

11 Lillian Hoddeson and Vicki Daitch, 2002, *True Genius: The Life and Science of John Bardeen*, Joseph Henry Press, Washington D.C., p. 122; Spencer Weart, 'Bight Ideas: The First Lasers', *American Institute of Physics*, http://history.aip.org/history/exhibits/laser/sections/whoinvented.html.

12 Lawrence F. Katz and Robert A. Margo, 2013, 'Technical Change and the Relative Demand for Skilled Labor: The United States in Historical Perspective', *The National Bureau of Economic Research*, Working Paper No. 18752, February.

13 Ian Stewart, Debapratim De and Alex Cole, Deloitte, 2015, 'Technology and People: The Great Job-Creating Machine', www2.deloitte.com/uk/en/pages/finance/articles/technology-and-people.html.

14 Michael Coelli and Jeff Borland, 2016, 'Job Polarisation and Earnings Inequality in Australia', http://fbe.unimelb.edu.au/__data/assets/pdf_file/0009/1427409/1192CoelliBorland.pdf, 24 February.

15 Rachel Levy and Matt Turner, 2016, 'One Brutal Chart from the Biggest Hedge Fund in the World Explains Everything', *Business Insider*, www.businessinsider.com.au/a-brutal-chart-from-bridgewater-explains-the-rise-of-trump-2016-11, 12 November.

16 Yoshua Bengio, 2016, 'Machines Who Learn', *Scientific American*, June, p. 46.

17 The difference between AI and AGI is that all AI systems today are designed to solve fairly specific tasks in relatively narrow domains. AGI would be a machine with general cognitive abilities, allowing it to perform almost any task that a human is capable of performing. AGI is also referred to as 'strong AI' or 'human-level machine intelligence'. A number of eminent researchers, including Stuart Russell and Nick Bostrom, have brought the risks of AGI to public attention. Russell and Bostrom note that AGI could potentially race right past human capabilities, to what Bostrom has labelled 'superintelligence'.

18 Stuart Russell and Peter Norvig, 2016, *Artificial Intelligence: A Modern Approach*, third edition, Pearson, Essex, pp. 3–4.

19 Evan Ackerman, 2016, 'SRI Spin-off Abundant Robotics Developing Autonomous Apple Vacuum', IEEE Spectrum, http://spectrum.ieee.org/automaton/robotics/industrial-robots/sri-spin-off-abundant-robotics-developing-autonomous-apple-vacuum, 15 August.

20 Dan Charles, 2015, 'Inside the Life of an Apple Picker', National Public Radio, www.npr.org/sections/thesalt/2015/10/23/448579214/inside-the-life-of-an-apple-picker, 23 October.

21 SRI International, 2016, 'Disrupting the Apple Cart: Abundant Robotics to Automate Orchard Harvests', www.sri.com/newsroom/press-releases/disrupting-apple-cart-abundant-robotics-automate-orchard-harvests, 10 August.

CHAPTER 3

1 'Nissan Is "Racing Forward" With Autonomous Cars, Says CEO Carlos Ghosn', 2015, *Forbes*, 27 October.

2 Intel News Release, 2016, 'BMW Group, Intel and Mobileye Team Up to Bring Fully Autonomous Driving to Streets by 2021', https://newsroom.intel.com/news-releases/intel-bmw-group-mobileye-autonomous-driving, 1 July.

3 The Boston Consulting Group, 2015, 'Back to the Future: The Road to Autonomous Driving: Selected Highlights for SlideShare', http://de.slideshare.net/TheBostonConsultingGroup/the-road-to-autonomous-driving, 8 January.

4 Steven Shladover, 2016, 'The Truth About 'Self-Driving' Cars', *Scientific American*, June.

5 Frey and Osborne, 2013, 'The Future of Employment'.

6 'Engines of Evidence: A Conversation with Judea Pearl', 2016, Edge Foundation, www.edge.org/conversation/judea_pearl-engines-of-evidence, 24 October.

7 Yuval Noah Harari, 2016, *Homo Deus: A Brief History of Tomorrow*, Harvill Secker, London, pp. 324–325.

8 Angus Knowles-Cutler, Carl Benedikt Frey and Michael A. Osborne, 2014, 'Agiletown: The Relentless March of Technology and London's Response', www2.deloitte.com/content/dam/Deloitte/uk/Documents/uk-futures/london-futures-agiletown.pdf.

9 The authors caution that the US and UK numbers are not directly comparable due to the differences in the detail of occupation classifications used.

10 Hugh Durrant-Whyte et al., 2015, 'The Impact of Computerisation and Automation on Future Employment'. Committee for Economic Development of Australia, Report on Australia's future workforce, http://adminpanel.ceda.com.au/FOLDERS/Service/Files/Documents/26792~Futureworkforce_June2015.pdf. June, p. 56.

11 Melanie Arntz, Terry Gregory and Ulrich Zierahn, 2016, 'The Risk of Automation for Jobs in OECD Countries: A Comparative Analysis', *OECD Social, Employment and Migration Working Papers*, No. 189, OECD Publishing, Paris, www.oecd-ilibrary.org/docserver/download/5jlz9h56dvq7-en.pdf, 16 June.

12 James Manyika, Michael Chui, Mehdi Miremadi, Jacques Bughin, Katy George, Paul Willmott and Martin Dewhurst, 2017, 'Harnessing Automation for a Future that Works', www.mckinsey.com/global-themes/digital-disruption/harnessing-automation-for-a-future-that-works, McKinsey & Company, January.

13 AlphaBeta, 2017, *The Automation Advantage*, Report, www.alphabeta.com/the-automation-advantage, Sydney, 8 August.

14 Australian Government, 2016, *Department of Employment, Industry Employment Projections*, 2016 Report, http://lmip.gov.au/default.aspx?LMIP/EmploymentProjections, March.

15 Central Intelligence Agency, *The World Factbook*, www.cia.gov/library/publications/the-world-factbook/fields/2012.html. See also Wikipedia, 2017, 'List of Countries by GDP Sector Composition', https://en.wikipedia.org/wiki/List_of_countries_by_GDP_sector_composition.

16 S.A. Hajkowicz et al., 2016, 'Tomorrow's Digitally Enabled Workforce: Megatrends and Scenarios for Jobs and Employment in Australia over the Coming Twenty Years', CSIRO, Brisbane.

17 Richard Susskind and Daniel Susskind, 2015, *The Future of the Professions: How Technology Will Transform the Work of Human Experts*, Oxford University Press, New York, p. 37, p. 301.

18 Aaron Smith and Janna Anderson, Pew Research Center, 2014, 'AI, Robotics, and the Future of Jobs', www.pewinternet.org/2014/08/06/future-of-jobs, August.

19 Ibid.

Chapter 4

1 Tyler Cowen, 2013, *Average is Over: Powering America Beyond the Age of the Great Stagnation*, Penguin Group, New York, pp. 4–5.

2 Oxfam, 2017, 'An Economy for the 99%: Australian Fact Sheet', www.oxfam.org/sites/www.oxfam.org/files/file_attachments/bp-economy-for-99-percent-160117-en.pdf, January, p. 2.

3 Brian Keeley, 2015, *Income Inequality: The Gap Between Rich and Poor*, OECD Publishing, Paris, p. 3.

4 Brynjolfsson and McAfee, 2014, *The Second Machine Age: Work, Progress, and Prosperity in a Time of Brilliant Technologies*, p. 129.
5 Oxfam, 2017, 'An Economy for the 99%', p. 1.
6 Andrew Leigh, 2017, 'Why Scott Morrison Isn't Entitled to His Own Facts on Inequality in Australia', *Business Insider*, www.businessinsider.com.au/why-scott-morrison-isnt-entitled-to-his-own-facts-on-inequality-in-australia-2017-7, 26 July.
7 Miles Corak, 2013, 'Income Inequality, Equality of Opportunity and Intergenerational Mobility', *Journal of Economic Perspectives*, Vol. 27, No. 3, p. 81. Note: Finding refers to earnings of fathers and sons because, as Corak puts it: 'It is not that studies of mothers, daughters, and the marriage market do not exist, only that father–son analyses are more common and permit a broader set of cross-country comparisons.'
8 Corak, *Journal of Economic Perspectives*, p. 80.
9 Silvia Mendolia and Peter Siminski, 2016, 'New Estimates of Intergenerational Mobility in Australia', *Economic Record*, Vol. 92, No. 298, p. 363.
10 OECD, 2010, 'A Family Affair: Intergenerational Social Mobility across OECD Countries', *Economic Policy Forums 2010: Going for Growth*, www.oecd.org/centrodemexico/medios/44582910.pdf.
11 Federico Cingano, 2014, 'Trends in Income Inequality and its Impact on Economic Growth', *OECD Social, Employment and Migration Working Papers*, No. 163, OECD Publishing, Paris, p. 18.
12 Era Dabla-Norris et al., 2015, *Causes and Consequences of Income Inequality: A Global Perspective*, International Monetary Fund, Washington, p. 19.
13 Andrew Leigh, 2015, 'Humans Need Not Apply: Will the Robot Economy Pit Entrepreneurship Against Equality?', speech, Fall 2015 Distinguished Public Policy Lecture, Northwestern University, 28 October.
14 Kaushik Basu, 2016, 'Is Technology Making Inequality Worse?', World Economic Forum, www.weforum.org/agenda/2016/01/is-technology-making-inequality-worse/, 6 January.
15 James Bessen, 2016, 'Computers Don't Kill Jobs But Do Increase Inequality', *Harvard Business Review*, 24 March.

16 World Bank, 2016, *World Development Report 2016: Digital Dividends*, World Bank, Washington, p. 120
17 Leigh, 2015, 'Humans Need Not Apply'.
18 OECD, 2013, 'Who's Smiling Now?', *OECD Observer*, http://oecdobserver.org/news/fullstory.php/aid/4227/Who_92s_smiling_now_.html.
19 Claudia Goldin and Lawrence Katz, 2009, *The Race Between Education and Technology*, Harvard University Press, Cambridge/London, pp. 287–323.
20 Goldin and Katz, *The Race Between Education and Technology*, p. 320.
21 Laurence Chandy and Kemal Derviş, 2016, 'Are Technology and Globalization Destined to Drive up Inequality?', Brookings, www.brookings.edu/research/are-technology-and-globalization-destined-to-drive-up-inequality, 5 October.
22 World Bank, 2016, *World Development Report 2016*, p. 118.
23 Yochai Benkler, 2016, 'What the World Bank Report on Tech-Related Income Inequality Is Missing', *The Guardian*, www.theguardian.com/technology/2016/jan/15/world-bank-income-inequality-report-silicon-valley, 15 January.
24 Chandy and Derviş, 2016, 'Are Technology and Globalization Destined to Drive up Inequality?'
25 World Bank, 2016, *World Development Report 2016*, p. 118.
26 Bessen, 2016, 'Computers Don't Kill Jobs But Do Increase Inequality'.
27 Philippe Aghion et al., 2015, 'Innovation and Top Income Inequality', *NBER Working Paper 21247*, Cambridge, p. 2. Note: Aghion and his colleagues point out that eleven of the fifty people who featured on the 2015 Forbes rich list were inventors.
28 Jaron Lanier, 2013, *Who Owns the Future*, Penguin Books, London, p. xx
29 Basu, 2016, 'Is Technology Making Inequality Worse?'
30 Jim Stanford, 2017, 'Labour Share of Australian GDP Hits All-Time Record Low', Centre for Future Work, briefing note, 13 June.
31 ACTU, 2013, 'A Shrinking Slice of the Pie', *ACTU Working Australia Paper*, pp. 2–3.
32 Greg Jericho, 2017, 'Australians Aren't Being Paid for their Productivity. Get Set for an Industrial Relations War', *The*

Guardian Australia, www.theguardian.com/business/grogonomics/2017/may/23/australians-arent-being-paid-for-their-productivity-get-set-for-an-industrial-relations-war, 23 May.

33 Chandy and Derviş, 2016, 'Are Technology and Globalization Destined to Drive up Inequality?'

34 Andrew Berg, Edward Buffie and Luis-Felipe Zanna, 2016, 'Robots, Growth and Inequality', *Finance & Development*, International Monetary Fund.

35 Some, like Albert Wagner from Union Square Ventures, have gone further to suggest wages could be cut to the point where it's cheaper to go with a new underclass than robots. As he puts it, 'There is little incentive to invest in a machine if it is in fact cheaper to have people do the work by hand.' See Albert Wagner, 2017, *Work After Capital: Labor*, GitBook, https://worldaftercapital.gitbooks.io/worldaftercapital/content/part-two/Labor.html.

36 Patrick Durkin, 2015, 'Robot Workers Cut Low-Skilled Jobs by Nearly Half', *The Australian Financial Review*, 1 October.

37 Edmund Phelps, 2013, 'Mass Flourishing: How it Was Won and Then Lost', *Reuters*, 16 August.

38 Paul Mason, 2015, *Postcapitalism: A Guide to Our Future*, Farrar, Straus and Giroux, New York, pp. 164–176.

39 Deloitte, 2017, 'The Deloitte Millenial Survey 2017', www2.deloitte.com/global/en/pages/about-deloitte/articles/millennialsurvey.html.

40 Barack Obama, 2017, 'Read the Full Transcript of President Obama's Farewell Speech', *Los Angeles Times*, 10 January.

41 Cowen, 2013, *Average is Over*, p. 5.

Chapter 5

1 Donald Horne, 2009 [1964], *The Lucky Country*, Penguin Books, Camberwell, Vic., p. 233.

2 Jim Chalmers, 2013, *Glory Daze: How a World-Beating Nation Got so Down on Itself*, Melbourne University Press, Carlton, Vic., pp. 75–77. For further reading, see also: Wayne Swan, 2014, *The Good Fight*, Allen & Unwin, Sydney.

3 Claire O'Neil and Tim Watts, 2015, *Two Futures: Australia at a Critical Moment*, Text Publishing, Melbourne, pp. 8–9, pp. 207–217.

4 George Megalogenis, 2016, *Balancing Act: Australia between Recession and Renewal*, Quarterly Essay 61, Black Inc., Carlton, Vic.

5 World Economic Forum, 2016, 'The Future of Jobs: Employment, Skills and Workforce Strategy for the Fourth Industrial Revolution', www.weforum.org/reports/the-future-of-jobs, 18 January.

6 S.A. Hajkowicz et al., 2016, 'Tomorrow's Digitally Enabled Workforce: Megatrends and Scenarios for Jobs and Employment in Australia over the Coming Twenty Years'.

7 Chris Mairs, 2014, UKForCE Submission to Maggie Philbin's Digitial Task Force, UK Forum for Computing Education, http://ukforce.org.uk/ukforce/ukforce-submission-to-maggie-philbins-digital-task-force, 11 July.

8 Report prepared by AlphaBeta for the Foundation for Young Australians, 2015, 'The New Work Order: Ensuring Young Australians Have Skills and Experience for the Jobs of the Future, Not the Past', www.fya.org.au/wp-content/uploads/2015/08/fya-future-of-work-report-final-lr.pdf.

9 This point was made by Jim and his Labor colleague Tim Watts in December 2014, when they observed that even primary school children have an appreciation of how important these skills will be to their future. For more information, see Jim Chalmers and Tim Watts, 2014, 'Kids Should Code: Why Computational Thinking Needs to Be Taught in Schools', *The Guardian*, www.theguardian.com/commentisfree/2014/dec/19/kids-should-code-why-computational-thinking-needs-to-be-taught-in-schools, 19 December.

10 Editorial, 2017, *Sydney Moring Herald*, 'Rewards in Hard Maths: Personally and Nationally', 31 January, www.smh.com.au/comment/smh-editorial/hard-maths-brings-rewards-personally-and-nationally-20170129-gu14b1.html.

11 Maaike Wienk, 2016, 'Discipline Profile of the Mathematical Sciences, 2016', Australian Mathematical Sciences Institute, http://amsi.org.au/publications/discipline-profile-mathematical-sciences-2016.

12 John Kennedy, Terry Lyons and Frances Quinn, 2014, 'The Continuing Decline of Science and Mathematics Enrolments in Australian High Schools', *Teaching Science*, Vol. 60, No. 2, June, https://eprints.qut.edu.au/73153/1/Continuing_decline_of_science_proof.pdf.

13 David G.W. Pitt, 2015, 'On the Scaling of NSW HSC Marks in Mathematics and Encouraging Higher Participation in Calculus-Based Courses', *Australian Journal of Education*, Vol. 59, No. 1, 10 February.

14 See also Pallavi Singhal, 2017, '"Significant Scaling Advantage": Why More HSC Students Are Opting for Lower-Level Maths', *The Sydney Morning Herald*, www.smh.com.au/national/education/significant-scaling-advantage-why-more-hsc-students-are-opting-for-lowerlevel-maths-20170515-gw505a.html, 15 May.

15 Andrew Leigh, 2015, 'Robots, Remuneration and Restructuring: How do Technology and Inequality Shape One Another, and What Should We Do About It?', Annual Sir Leslie Melville Lecture, Australian National University, November.

16 Kate Jones, 2017, 'STEM Specialists and IT Support for State Schools', Queensland Government media release, 23 June.

17 Suzanne Straw, Oliver Quinlan, Jennie Harland and Mathew Walker, 2015, 'Flipped Learning Research Report', National Foundation for Educational Research (NFER) and Nesta, www.nfer.ac.uk/publications/NESM01/NESM01.pdf.

18 Lexy Hamilton-Smith, 2017, 'Learning Curve; Coding Classes to Become Mandatory in Queensland Schools', ABC News, www.abc.net.au/news/2016-11-17/coding-classes-in-queensland-schools-mandatory-from-2017/8018178, 6 February.

19 Federal Labor announced a comprehensive policy for coding in schools ahead of the 2016 election. See Bill Shorten, 2015, 'Labor's Plan for Coding in Schools', www.billshorten.com.au/labors-plan-for-coding-in-schools, 18 May.

20 Jeannette M. Wing, 2012, 'Computational Thinking', Microsoft Asia Faculty Summit, http://research-srv.microsoft.com/en-us/um/redmond/events/asiafacsum2012/day1/Jeannette_Wing.pdf, 26 October.

21 Stephen Wolfram, 2016, 'How to Teach Computational Thinking', Stephan Wolfram Blog, http://blog.stephenwolfram.com/2016/09/how-to-teach-computational-thinking, 7 September.

22 International Society for Technology in Education and the Computer Science Teachers Association, 2011, 'Operational Definition of Computational Thinking for K-12 Education', www.iste.org/docs/ct-documents/computational-thinking-operational-definition-flyer.pdf?sfvrsn=2.

23 Donald E. Knuth in Cristopher Moore and Stephan Mertens (eds), 2011, *The Nature of Computation*, Oxford University Press, New York, p. 41.

24 D. Weintrop et al., 2016, 'Defining Computational Thinking for Mathematics and Science Classrooms.' *Journal of Science Education and Technology*, Vol. 25, No. 1, pp. 127–147.

25 Frank Levy and Richard Murnane, 2013, 'Dancing with Robots: Human Skills for Computerized Work', 2013, https://dusp.mit.edu/uis/publication/dancing-robots-human-skills-computerized-work, 1 June.

26 Miles Corak, 2013, 'Income Inequality, Equality of Opportunity, and Intergenerational Mobility', Forschungsinstitut zur Zukunft der Arbeit, IZA DP No. 7520, http://ftp.iza.org/dp7520.pdf, July.

27 Michelle Y. Simmons, 2017, 'Australia Day address', 24 January, www.australiaday.com.au/events/australia-day-address/2017-speaker-professor-michelle-y-simmons/.

Chapter 6

1 Edgar Royston Pike, 1966, *Human Documents of the Industrial Revolution in Britain*, Routledge, London/New York, pp. 121–122.

2 Building on comments from Larry Summers, in Ignatieff, 2014, 'We Need a New Bismarck to Tame the Machines'.

3 Kentaro Toyama, 2015, *Geek Heresy: Rescuing Social Change from the Cult of Technology*, 'Description', https://geekheresy.org.

4 To be clear, our notion of technological trickle-down is close, but different in meaning to 'trickle-down techonomics', a phrase coined by software developer Jon Gosier. He describes it as the idea of blindly championing technology which often only helps a select few at the top, with the expectation that everyone in society will eventually benefit from it. We focus largely on the distribution of the economic and workplace impacts of technology, rather than

the individual innovations themselves. See Jon Gosier, 2015, 'The Problem with "Trickledown Techonomics"', speech, TED, www.ted.com/talks/jon_gosier_the_problem_with_trickle_down_techonomics/transcript?language=en, March.

5 Nicholas Carr, 2013, 'All Can Be Lost: The Risk of Putting Our Knowledge in the Hands of Machines', *Atlantic*, www.theatlantic.com/magazine/archive/2013/11/the-great-forgetting/309516, November.

6 Nicholas Carr, 2015, *The Glass Cage: Who Needs Humans Anyway?*, Vintage, London, p. 227.

7 Peter Fisher, 2017, 'When Things Go Wrong in an Automated World, Would We Still Know What to Do?', *The Conversation*, 20 March.

8 Jason Furman et al., 2016, 'Artificial Intelligence, Automation and the Economy', Executive Office of the President, 20 December.

9 Kinley Salmon, 2015, 'Education and Training – Lessons from Denmark', The Future of Work Commission, New Zealand, 2 December.

10 Ministry of Foreign Affairs of Denmark, 2017, 'Flexicurity', *Denmark.dk: The Official Website of Denmark*, http://denmark.dk/en/society/welfare/flexicurity.

11 AlphaBeta, 'Automation and the Future of Work'

12 James Manyika et al., 2017, 'Harnessing Automation for a Future that Works', www.mckinsey.com/global-themes/digital-disruption/harnessing-automation-for-a-future-that-works, McKinsey & Company, January.

13 Catherine Clifford, 2016, 'Elon Musk: Robots Will Take Your Jobs, Government Will Have to Pay Your Wage', *CNBC.com*, www.cnbc.com/2016/11/04/elon-musk-robots-will-take-your-jobs-government-will-have-to-pay-your-wage.html, 4 November.

14 Martin Ford, 2015, *The Rise of the Robots: Technology and the Threat of a Jobless Future*, Basic Books, New York, pp. 257–261.

15 Peter Whiteford, 2014, 'The Budget, Fairness and Class Warfare', *Inside Story*, http://insidestory.org.au/the-budget-fairness-and-class-warfare, 5 August.

16 Andrew Leigh, 2017, 'How Can We Reduce Inequality?', speech, Crawford School of Public Policy, Canberra, 20 April.

17 Peter Hartcher, 2016, 'What if Everyone Were Given Money for Nothing?' *The Sydney Morning Herald*, 7 June.

18 The Australasian Faculty of Occupational and Environmental Medicine, 2011, 'Australian and New Zealand Consensus Statement on the Health Benefits of Work', Royal Australasian College of Physicians, Sydney, October, p. 13.

19 Ross Dawson, 2017, 'Shaping a Positive World as We Move into a "Post-Work Economy"', RossDawson.com, http://rossdawson.com/blog/shaping-positive-world-move-post-work-economy, 4 January.

20 Judith Bessant, 2000, 'Civil Conscription or Reciprocal Obligation: The Ethics of Work for the Dole', *The Australian Journal of Social Issues*, Vol. 35, No. 1, p. 26.

21 Heath Aston, 2016, 'Work for the Dole Has Little Effect on Finding Work: Review', *The Sydney Morning Herald*, 11 February.

22 McCrindle, 2014, 'Job Mobility in Australia', The McCrindle Blog, http://mccrindle.com.au/about-Australias-social-researchers, 18 June.

23 *The Economist*, 2014, 'The Three Types of Unemployment', www.economist.com/blogs/economist-explains/2014/08/economist-explains-8, 18 August.

24 Martin Elliott Jaffe, 2007, 'Spiral Careers Support the 21st Century Workforce', National Career Development Association, www.associationdatabase.com/aws/NCDA/pt/sd/news_article/5407/_PARENT/layout_details/false, 11 January. See also Peter Guber, 2014, 'Why the Career Ladder no Longer Matters', World Economic Forum, www.weforum.org/agenda/2014/10/peter-guber-career-ladder-skills, 16 October.

25 *The Economist*, 2017, 'Retraining Low-Skilled Workers', 12 January.

26 Chong Zi Liang, 2015, 'Singapore Budget 2015: Every Singaporean Above 25 to get $500 for a Start Under SkillsFuture', *The Straits Times*, 23 February.

27 Benkler, 2016, 'What the World Bank Report on Tech-related Income Inequality Is Missing'.

28 There have been some limited positive developments in addressing workers' entitlements and protections in the 'gig economy', but it's still an issue. See Anna Patty, 2017, 'Airtasker and Unions Make Landmark Agreement to

Improve Pay Rates and Conditions', *The Sydney Morning Herald*, 1 May.

29 Nick Hanauer and David Rolf, 2017, 'This New Social Security System Solves the Problem of the Gig Economy', *Evonomics*, http://evonomics.com/new-social-security-system-sharing-economy-hanauer, 10 March.

30 James Pawluk, undated, 'Wage Administration', draft paper.

31 For Labor's policy to address issues in the sharing economy, see Australian Labor Party, 2016, 'National Sharing Economy Principles', www.alp.org.au/sharingeconomy.

32 Alison Griswold, 2016, 'The Author of "The Sharing Economy" on Uber, China, and the Future of Work', *Quartz*, https://qz.com/710515/arun-on-sharing-economy, 21 June.

33 Benkler, 2016, 'What the World Bank Report on Tech-related Income Inequality Is Missing'.

34 Trebor Scholz, 2017, 'Platform Cooperativism vs the Sharing Economy', *Medium*, 5 December.

35 Yolanda Redrup, 2016, 'Top Investors Demand Changes to Australian Employee Share Scheme Rules', *The Australian Financial Review*, 9 February. American politicians Mark Warner and Mitch Daniels' work at the Aspen Institute's Future of Work Initiative focuses on more employee shares options and a minimum holding period. The institute's work focuses on policies that, among other things, 'encourage businesses to make greater investments in workers' and encouraging long-term value creation by 'promoting expanded participation in corporate equity to democratize ownership'. See Elliott Gerson, 2016, 'The Future of Work', The Aspen Institute, www.aspeninstitute.org/blog-posts/the-future-of-work, 20 July.

36 Sean Farrell, 2016, 'Workers on Boards: The Idea is Not Going Away', *The Guardian*, www.theguardian.com/business/2016/oct/02/theresa-may-tory-conference-workers-on-boards, 2 October. See also Christopher Williams, 2016, 'Theresa May Backtracks on Putting Workers on Company Boards', *The Telegraph*, 21 November.

37 Robert Shiller, 2017, 'Why Robots Should be Taxed if They Take People's Jobs', *The Guardian*, www.theguardian.com/business/2017/mar/22/robots-tax-bill-gates-income-inequality, 22 March.

38 Robert Atkinson, 2017, 'In Defense of Robots: Why We Should Not Reject Technology in Order to "Protect" Workers', *National Review*, 21 April.

39 This point about productivity gains is the subject of Shawn Donnan, 2017, 'Global Productivity Slowdown Risks Social Turmoil, IMF Warns', *The Financial Times*, 4 April.

40 Lawrence Summers, 2017, 'Robots Are Wealth Creators and Taxing Them Is Illogical', *The Financial Times*, 6 March. Partly for these reasons, the concept was ultimately rejected by the EU parliament in early 2017. Luxembourg MEP Mady Delvaux wrote: 'Bearing in mind the effects that the development and deployment of robotics and AI might have on employment and, consequently, on the viability of the social security systems of the Member States, consideration should be given to the possible need to introduce corporate reporting requirements on the extent and proportion of the contribution of robotics and AI to the economic results of a company for the purpose of taxation and social security contributions.' See Mady Delvaux, 2016, 'Draft Report with recommendations to the Commission on Civil Law Rules on Robotics', European Parliament Committee on Legal Affairs, No. 2015/2103, Brussels, p. 10.

41 Furman et al., 2016, 'Artificial Intelligence, Automation and the Economy', p. 44.

42 2016, 'Measuring Economies: The Trouble with GDP', *The Economist*, 30 April.

43 Clive Hamilton and Hugh Saddler, 1997, 'The Genuine Progress Indicator: A New Index of Changes in Well-Being in Australia', Australian Institute, Discussion Paper Number 14, October.

44 Chris Bowen, 2016, 'The Case for Opportunity', speech, address to the Chifley Research Centre, 22 November.

45 Tim Watts has argued this in his proposal to adopt a version of economist Nicholas Gruen's National Information Policy, which includes establishing an Independent Data Council modelled on organisations in New Zealand and the UK that include experts from government, academia and the private sector. See Australian Labor Party, 2016, 'National Information Policy', www.alp.org.au/nationalinformationpolicy.

46 Peter Harris, 2017, 'Data - The Thing that Ties it All Together', speech, Address to the Committee for Economic Development of Australia, Melbourne, 22 March.

47 See Tim Watts, 2015, 'Look at the Big Picture: There's Money in Data', *The Australian*, 7 December.

48 Roger Burkhardt and Colin Bradford, 2017, 'Addressing the Accelerating Labor Market Dislocation from Digitalization', policy brief, *The Brookings Institution*, March, p. 5.

CHAPTER 7

1 Jim Chalmers, 2015, 'Service Exports Key to China', http://jimchalmers.org/Media/Opinion-Pieces/Post/4785/Service-Exports-Key-to-China, 19 October.

2 RMIT-ABC Fact Check, 2015, 'Fact Check: How Many Students Leave Higher Education with STEM Qualifications', www.abc.net.au/news/2015-10-23/fact-check-bill-shorten-stem-qualifications-australia/6828470, 23 October; DrEducation: Global Higher Education Research and Consulting, 'Statistics on Indian Higher Education 2012–2103', www.dreducation.com/2013/08/data-statistics-india-student-college.html.

3 The Global Innovation Index 2016, www.globalinnovationindex.org/gii-2016-report.

4 2017, 'Retraining Low-skilled Workers', *The Economist*, www.economist.com/news/special-report/21714175-systems-continuous-reskilling-threaten-buttress-inequality-retraining-low-skilled, 12 January.

5 Victor Lavy and Edith Sand, 2015, 'On the Origins of Gender Human Capital Gaps: Short and Long Term Consequences of Teachers' Stereotypical Biases', The National Bureau of Economic Research, www.nber.org/papers/w20909, January.

6 Steve Lohr, 2015, 'Stephen Wolfram Aims to Democratize His Software', *The New York Times*, https://bits.blogs.nytimes.com/2015/12/14/stephen-wolfram-seeks-to-democratize-his-software/?smid=tw-nytimesbits&smtyp=cur&_r=1, 14 December.

7 For details on how to participate, see http://hippyaustralia.bsl.org.au/about.

8 A. Norton and B. Cakitaki, 2016, 'Mapping Australian Higher Education 2016', Grattan Institute.

9 Megan Beck and Barry Libert, 2017, 'The Rise of AI Makes Emotional Intelligence More Important', https://hbr.org/2017/02/the-rise-of-ai-makes-emotional-intelligence-more-important, 15 February.